Earth Etudes for Elul

Spiritual Reflections for the Season

Edited by Katy Z. Allen

Strong Voices

Medfield, MA

Strong Voices Publishing
P.O. Box 731
Medfield, MA 02052
www.strongvoicespublishing.com
"from heart to mind—we cross bridges"

Credits: Cover design by Shirley Riga. Front and back cover photos by Gabriele Mezger. Title font is Aldolphus Serif.

All essays and poems in this volume were originally published on the blog of Ma'yan Tikvah—A Wellspring of Hope (www.mayantikvah.blogspot.com) and on Jewcology.org.

Sarah Chandler, "Ready for Withering Flowers" published in *The New Farmer's Almanac, Volume III: Commons of Sky, Knowledge, Land, Water*, Greenhorns, 2017

David Krantz, "Elul: The Month for Climate Action" Copyright © 2015-2018 David Krantz, who is supported by an IGERT-SUN fellowship funded by the National Science Foundation (Award 1144616).

ISBN: 978-0-9905361-6-1
eISBN: 978-0-9905361-5-4
Library of Congress Control Number: 2018908688
Revised October 2018

Introduction

The month of Elul in the Jewish calendar leads us up to the first day of the new year, Rosh HaShanah, and the ensuing *Yamim Noraim*, Days of Awe. Elul generally begins in August. As summer wanes, the nights come earlier with Shabbat evenings growing longer and days shorter. The early evenings and cool nights convey a sense of urgency, for we know in our minds, even if not our hearts, that summer cannot go on forever, and the crisp days of autumn are not far away. We hear the call of the shofar calling us to begin to prepare, to unpack, to unearth, and to explore, that we may be ready to engage with the One, the Mystery, the Spirit, on the holiest days of the year.

Elul is a time for reflection, a time for *t'shuvah*, of turning and re-turning to G!d and to our best selves, in preparation for the Days of Awe. It is time for *heshbon hanefesh*, examining our hearts and souls. We may ask ourselves: What have I done? What do I wish I had done? How have I changed? What do I hope to be and do in the future? How did I impact my loved ones? The world? How do I want to impact them? Elul is the time for us to begin to make atonement for the things we wish we had or hadn't done, and to renew ourselves, to do all we can to get ourselves to change.

Elul is a time to turn away from the ways in which we have missed the mark, to make restitution as needed, and to return to our best selves. It is a time to be reborn, transformed, and renewed. It is also a time of love and caring. The letters of the month's name come from the verse, "*Ani l'dodi, v'dodi li*–I am my beloved's, and my beloved is mine" (Song of Songs 6:3). Elul is a time of building better relationships with our beloveds–with each other, with G!d, and with the Earth.

In addition to the timeless questions, new ones, not previously required for us, also arise: What does *t'shuvah* mean for us in this era of climate disruption and environmental degradation? How do we become the best we need to be for humanity and for our planet, within the confines of our physical being and the social structures of our modern world?

There are no easy answers to these questions. To help you on this journey, you will find in this volume a series of reflections for the month of Elul, *divrei* Earth—teachings that connect Earth and Torah called *Earth Etudes for Elul*. You may ask, why "Earth Etudes for Elul"? Especially, why "etudes"?

An etude is a short musical exercise, generally for one instrument, that is meant to improve one's techniques on the instrument or to demonstrate one's skill. Each "etude" here is a poem or essay that connects in some way to our beloved planet Earth and has some kind of reflection about *t'shuvah*. They remind us that we cannot disconnect ourselves from all that surrounds us and that we are part of an intertwined whole so incredibly diverse, rich and amazing. They remind us we are not alone.

These reflections on *t'shuvah* and Earth by a variety of rabbis, environmentalists, gardeners, poets, and other thoughtful people, reflect many points of view and ways of looking at the world and the process of *t'shuvah*. Originally published on the blog of Ma'yan Tikvah—A Wellspring of Hope (www.mayantikvah.blogspot.com) and on Jewcology.org, these "etudes" are intended to provide food for thought during a sacred time of year, the time of the turning of the seasons and the turning of our calendars, and thus to help us improve our proficiency and become better people, during Elul and throughout the year.

Thus, the three-fold alliteration of *Earth Etudes for Elul*.

In this collection, you will consistently find the Earth-*t'shuvah* connection, but beyond that you will experience a wide variety of styles of writing and of presentation, as well as a range of understandings and thoughts about *t'shuvah* and the Divine. Your invitation is to dip in and sample. May the diversity inspire you.

May your journey through Elul, the Days of Awe, and life in general be rich, meaningful, and fulfilling, and may you find new wisdom, strength, and courage. May you discover within yourself previously untapped resources for confronting the challenges of life, and may you experience new and unexpected ways of being in relationship to yourself, your loved ones, the Earth, and the One Source of All.

> *Kol tuv*—all the best,
> Katy Z. Allen
> July 2018 / Tammuz 5778
> Wayland, Massachusetts

Acknowledgements

First and foremost, I would like to thank all the writers who have given of their time, their creativity, and their wisdom to help the Earth Etudes for Elul become an increasingly rich resource over the years. This project, both annually online and in this book, would be impossible without them.

I would like to thank my spouse, Gabi Mezger, who has unfailingly supported me in this and so many other projects over the years. Her support and belief in my visions is so important to my journey.

I would like to thank Carol Reiman for helping me get started on this project by reading through many of the etudes and later for proofreading it, and Rabbi Peg Kershenbaum for proofreading as well. I am extremely grateful to Thea Iberall and Shirley Riga for their support and for their editorial and publishing help. I would like to thank Strong Voices Publishing for their interest in this project, particularly because our missions align.

Table of Contents

Elul 1

Alarm Clock for Our Souls

Katy Z. Allen

It's Elul.
Once again.

We'll hear the shofar in the mornings
trying to wake us up.
An alarm clock for our souls.

For every morning, the Sun rises.

Sometimes I'd like to hit snooze,
But *Ruach HaKodesh*—the Holy Spirit—
won't let me,
or,
if S/He/It does,
the interval isn't long
until the alarm clock rings once more.
I have no choice.
I must keep moving forward.

For every evening, the Sun sets.

But there is, at times, joy in that, too,
in the moving forward—
unimaginable, overwhelming, excruciating joy.
For forward means letting go of pain,
hurt,
tenderness—
sometimes slowly,
and sometimes in huge glorious leaps
that leave me feeling so peaceful
I wonder if life is real.

And it is.

For every morning, the Sun rises.

Yes, hurt comes again,
and with it anger.
But growth keeps happening.
And when I do hit snooze,
the interval
until the next shofar blast
is shorter
and wakes me up more
than before,
to rediscover the joy
and the peace.

For every evening, the Sun sets.

From the Perspective of the Ninth of Av

Shoshana Brown

Writing on the mourning day of Tisha B'Av, I am inclined to think of this "etude" as rather more of a *kinah* (lament) for the magnificent temple of our Earth, third planet in our solar system. Not to say that Earth is a *churban*, a ruin, like our ancient Temple in Jerusalem, but to say that like that once-beating spiritual heart and ritual nerve-center of the nation of Israel, our planet is both magnificent and utterly vulnerable to the predations of human greed, violence, and recklessness.

And yet, I have turned the analogy inside-out, for it was the Temple that was built to mirror the grandeur of Creation, with its seven-branched menorah symbolizing Creation's seven days and shaped like almond branches, its cedar wall-carvings of palm trees and flowers, its two great bronze pillars ornamented with pomegranate patterns (perhaps symbolizing the Tree of Knowledge and the Tree of Life), its bronze basin in the courtyard called *Yam* (Sea), and for burnt offerings, its bronze altar which may have been experienced as a kind of "micro-Sun." Humans could not be perfect stewards of Eden and its surroundings, and so a system of rituals in a "micro-Eden" was established, a place where humans could come and seek atonement, ask forgiveness for their failings, and experience the immanence of God that the first man and woman experienced in Eden where they could hear "the sound of God" walking amongst them in the cool of the day.

Apparently the *Kohanim* (priests), the *Levi'im* (Levites), and the monarchy (Solomon oversaw the building of the First Temple) imagined that they would be better guardians of this micro-Eden than they were of the macrocosm. But greed, lust for power, political intrigue—all the usual suspects—led to the end of the United Monarchy (ca. 930 BCE) and the exodus of many of its priests, prophets, Levites, and members of the royal house to the North to set up the separate kingdom of Israel, events that would culminate in the eventual destruction of both kingdoms and the burning of Solomon's Temple by the Babylonians in 586 BCE.

So much for trying to perfect the world by theurgy! Meanwhile, Planet Earth continues her life, though battered by human exploitation and pollution of her air, soil and waters, which nevertheless in some places are healing, and in others becoming devastated beyond repair. We recite every morning in the blessing before the *Sh'ma*: "You illumine the Earth and its inhabitants with compassion; in Your goodness You renew, day after day and continually, the works of creation. How varied are Your works, Adonai! With wisdom You created them all—the Earth abounds with Your creations!"

God renews, day after day and continually; every day is a re-creation. And every day we must strive to attune ourselves to both Creator and Creation so that we do not become destroyers of Eden/the Temple/Creation again. *T'shuvah* requires an acknowledgement of our sins, a feeling of remorse, and some concrete plans to do better going forward. How might we do that? Let us turn to attune ourselves to the holiness of creation, let us re-turn to "Eden," immersing ourselves regularly in nature where we can experience God more immanantly than in our worlds of bricks and mortar or cyberspace, and let us seek out ways to become both guardians of Creation and partners with God, renewing Creation day after day.

I Am My Beloved and My Beloved is Me

Daniel Kieval

> "I have a friend who reads people's auras. He sees all sorts of colors like green & red & purple. He says anyone can do it. All it takes is forgetting everything you think you know & just looking. I've tried it & even though I haven't seen any colors yet, everyone I meet looks so beautiful when I stop knowing everything, that it's pretty hard to go back to the old way."[1]

---

[1] Andreas, Brian. "Beautiful People." www.storypeople.com/2013/12/16/beautiful-people. Accessed 2 Sept 2013.

3

Such is the mysterious beauty of our world that when we observe any part of it deeply we have no choice but to fall in love.

Many naturalists and nature educators will say that the best way to develop a connection with the Earth is to practice what is called a "Sit Spot." Here's how it works: Choose one place in the world and spend time there daily, at all times of day and night, in all weather, in all seasons. In your spot, sit in silence and focus fully on the world around you. As you learn to quiet your mind and let go of everything you think you know, you become open to receiving what nature is presenting to you in that moment. Over time you gain a deep sense of the patterns of life around your Sit Spot and, just maybe, you fall in love.

A personal connection with Earth is not something new we have to acquire. Every one of us has carried it in our bodies since the first *Adam* (human) was formed from the *Adam-ah* (Earth). By turning all of our awareness to nature's gifts, we come home again to that relationship which we've had all along.

In Elul, we focus on the process of *t'shuvah*—returning, coming home— through personal reflection and examination. What the Sit Spot is to the Earth, *t'shuvah* is to our own souls. We visit our "Sit Spot of the Self" daily; we see what it's like there in all weather and moods. We let go of what we think we know about ourselves and instead we quiet down and listen. We discover the subtle beauties of our inner ecology.

Our souls, like the Earth, have always been there waiting for us, but we lose touch with them as the clutter of everyday life fills up our heads. In Elul, we visit our souls with devotion until we fall in love with ourselves again. That is what it means to do *t'shuvah*: to come back to our pure essential nature that is as unspoiled and good and true as every other primordial piece of Creation. Only after we've done this are we ready to face the infinite on Rosh Hashanah and Yom Kippur.

Spend time with yourself this Elul. Be quiet. Be curious. Be present. Let go of judgment and observe openly and honestly. In so doing, may you come home again to a loving relationship with the wild Earth and your own perfect soul.

Elul 2

Keeping Agreements as Spiritual Practice

David Jaffe

I am a people pleaser. On the surface that may sound just fine. I get along well with people, care about people and want to give them what they want. But the motivations for my people pleasing reveal its dark underside. I don't like conflict, so I will do whatever is necessary to make sure people like me. For example, I will say yes to things I know I will never do, sacrificing my integrity to avoid the momentary discomfort and hard feelings of saying no.

I am not sharing this publicly to self-flagellate. Rather, my own condition is instructive for many people because these patterns of behavior are not of my own invention and are not a "personal problem." On the contrary, I have been trained well by the dominant Protestant middle-class culture of the United States to be a good cog in the capitalist machine. This training teaches one to work hard, keep your head down, conform, avoid conflict, and get people to like you, with the guarantee that one will achieve a level of comfort that is the goal of life. Then, do whatever possible not to lose this comfort. This means avoiding hard moments with other people that might create conflict.

Caitlin Breedlove, a community organizer with the Auburn Seminary, names the broader implications of conflict avoidance in Elizabeth Aeschlimann's powerful master's thesis.[1] Breedlove, who was raised in a working-class family, recounts numerous experiences with middle-class college students and organizers who said yes to certain agreements and then broke their word in the course of working together. The difficult changes community organizing seeks to make take relationships that people can count on. Without knowing that someone really has your back, it is hard to fight for real change. Breedlove tells Aeschlimann that people involved in campaigns, "…really wanted spiritual accompaniment on the road…I think when you're really accompanying them, you have a spiritual covenant with them. You've given them your word, and you've asked something in return." This idea of covenant is key for Breedlove. Covenanting with someone means that you are there for that person in a real and continuous way and will not break your agreements, even if it makes you deeply uncomfortable.

---

[1] Aeschlimann, Elizabeth. *Getting Mixed Up With Each Other* (Unpublished master's thesis). Harvard Divinity School, Cambridge, MA, 2017.

Am I willing to give up the momentary comfort of saying yes to a request and instead have the integrity to make agreements I will keep? Am I willing to always keep my word? A lot is at stake in the white middle class in our country confronting these questions. On an environmental level, we have an implicit agreement with future generations to steward the Earth and leave it in better shape than we found it. Will we follow through with that agreement? Will we keep our word to our children or sacrifice our integrity for short term economic comfort? In the language of *Middot* (Jewish soul traits), the commitment and integrity that Breedlove advocates is called *Emunah* (trustworthiness and reliability). A traditional blessing given to couples upon marriage is, "May you build a *bayit ne'eman b'Yisrael* (a trustworthy and reliable home)."

This Elul, I am asking myself to sacrifice comfort and risk creating conflict with people by only saying yes to things I can actually do. Taking on a commitment, no matter how small, means actually following through and doing it. If everyone with these same people-pleasing patterns can commit to keeping our word, we can make this world a *bayit ne'eman*, a reliable, trustworthy home, where we responsibly steward this miraculous Earth for the generations to come.

Water Down the Wall

Carol C. Reiman

Borne
on the water
that etches
the rock
(of tablets
and of temples),
the breath of life
glistens
as it falls
down
and down
and
down
the wall,
pooling below, in blue green deep,
a balm for bathers, to wash away
the ash of grief and tired day;

turning on,
the waters pass slowly here,
clotted
by the blood of battle,
iron arms lapped
in reeds—
land of dust and stone
sealed off to some—
left in pieces,
separated parts.
Eicha![1]
From the voices
of those gathered,
tuned to the shofar,
condensing
breath
of all with open ear and heart,
a mist of spray
forms,
rippling out,
spiraling,
to rise,
amidst the mix,
until
the
water
once again
glistens
with its light;
borne upon
its journey,
making
its way
down
the
wall.

---

[1] Woe!

Beginnings and *T'shuvah*

Howard Smith

For six weeks beginning in Elul, Jews engage in introspection. "Elul," taken as an acronym: I my beloved's and my beloved is mine, introduces the principle that *t'shuvah* is about healing relationships: with self, others, God, and as well, with the Earth. The season of *t'shuvah* is not completed until we return in the Torah cycle to the beginning, *Beresheit* (Genesis), the story of Creation.

Understanding and coming to grips with the physical Universe in which we live is an integral part of the process of *t'shuvah*. It gives us perspective on who we are, where we came from, where we are going, and God's plan. This is why the Torah begins with what is, as Rashi astutely notes, a seemingly irrelevant story. If we cannot appreciate the true nature of the Earth and cosmos in which we live, then we cannot truly understand the web of relationships and obligations that bind us.

About 13.7 billion years ago, the Universe as we know it began from an infinitesimally small point that exploded in a creative event that scientists call "the big bang." It has been expanding and evolving ever since. That realization was the cumulative result of decades of meticulous measurements and mathematical calculations that culminated with Edwin Hubble's 1929 observation that other galaxies were systematically moving away from our own Milky Way galaxy, in accord with Einstein's then recent theory of relativity.

About four hundred years earlier, the Kabbalists of Tzefat had offered their own perspective on the Torah's lessons of beginning. They weave an intricate account of an expanding Universe, layers upon layers of light emanating from a primal point, but with a reason beyond the simple physics: humanity has a role to play in this cosmic drama. *Tikkun olam* is humanity's task—to heal the breaches and injustices of our society, imperfections that were reflected in the very fabric of the newly formed cosmos. Caring for the Earth is one of these tasks. *G'mar Chatimah Tovah*. May you be sealed for goodness.

Elul 3

Holy Limits

David Arfa

> "Self-respect is the root of discipline: The sense of dignity grows with the ability to say no to oneself."[1]—Rabbi Abraham Joshua Heschel

When you were a kid, did you also feel the unbounded delight that I felt when I learned that a few small coins could get me individually wrapped yellow cakes with creamy filling? In my childhood, I felt that this unbounded joy was the ultimate freedom.

Growing up, I learned that instant gratification has grave impacts on our lives, our communities and the entire planet. Thanks to corporations built on historic conquest, slavery, massive waste and pollution, we have expanded to the point where we can carry all the musicians of the world in our pocket; we can place all the foods of the world on our table; we can search through all the hard-won intellectual treasures of the world; and we can play and watch all the games, movies, comedians and news of the world on demand. For the first time, we can buy stuff at the push of a button.

And yet, we all know our planet is in dire need, waiting for us to collectively learn limit-setting. Our planet is being severely impacted by our collective means of energy production, food systems, transportation choices and buying habits. Our planet is changing and human society cannot withstand continued consumer growth. We need to recognize limits. We need to learn to live with the reality of riddles like this: lily pads are doubling on a pond every day, Day 1—1, Day 2—2, Day 3—4, Day 4—8 and so on. On day 30, the pond is filled. On what day is the pond half filled?[2]

Let's dive into our treasury of stories and see what may be imagined. I am writing one week after the destruction that is Tisha B'Av. It is now the time of consolation. And yet, in *Torah, Moshe Rabbeinu*, Moses our great teacher, is begging to cross over into the good land. He is old, but still of

---

[1] Heschel, Abraham Joshua. *God in Search of Man: a Philosophy of Judaism*. Farrar, Strauss and Giroux, 1955, p. 216. While not the exact quote, I believe this may be the source for this often-cited quote.

[2] Riddle Answer: The 29th day. And the 28th day the pond is only a quarter filled. The 27th day one-eighth. The 26th day? From Meadows, Donella. H., et al. *The Limits to Growth: A Report for the Club of Rome's Project on the Predicament of Mankind*. New York: Universe Books, 1972.

strong mind and strong eye. He has done so much: he grew up as Egyptian royalty; grappled with social injustice; responded with mortal violence; healed by land, love and a new wilderness community; followed his calling into a fateful showdown with Pharaoh and chose to spend the rest of his days leading an enslaved people into freedom. He is not a man used to taking no for an answer.

But here he is, reduced to pleading to be able to join his people as they cross over the river Jordan and enter the Land. Confusingly, the response from Gd is not consolation; instead it is the harsh rebuke—"Enough!" (Deuteronomy 3:26). I can't help but wonder, might there be consoling insights hidden for us?

My mind connects back to another 'Enough!' that was shouted by the Holy One to the expanding World at the very beginning of time. The Holy One noticed the speed of the expanding world and understood that if the expansion continued at break-neck speed, it would all be over in an instant. In response, the Holy One shouted 'Enough!' (*BT Hagigah 12a*). This slowing of time brought the gift of Shabbat into our world. This allowed us to slow down, remember what is most important in our lives and connect with the holiness at the root of our world.[1]

I wonder, might the Holy One saying 'Enough!', preventing Moses from crossing over the Jordan, also have gifted the rest of us with the opportunity to seek out and connect with holiness? After all, the Kabbalists say that if Moses actually crossed over with that desert generation, then he would have unified time and space, bringing the Messianic era and consequently ending the world as we know it.

Connecting these two 'Enough's!' allows us to imagine that these 'limits to growth' may be mythically woven into the very fabric of our world.

Amazingly, the Rabbis, at their mythopoetic finest, show us how to activate this power of limit-setting. They remind us that we are not just passive receivers, but also wielders of power as co-creators with the Holy One!

The Rabbis teach us that each week, when we lift the goblet of grape for Kiddush and we invoke words from Torah, we are holding and wielding the power that limits the expanding world. At that moment, we become partners with the Holy One when we say "*Vayechulu*, and they were finished" (Heaven and Earth were finished). The Midrashists creatively shift a few vowels, raising our participation by turning *Vayechulu* into *Vayechalu*: 'They finished' (Creator and Human finished together) (*BT Shabbat 119a*).

We all know how hard it is to set and maintain limits. Any of us who have tried a diet or negotiated the limits of screen time with children know

---

[1] Epstein, Kalonymus K. *Letters of Light: Passages from Ma'or va-shemesh.* Translated by Aryeh Wineman, Pickwick Publishers, 2015, p 124.

this; let alone limiting the work of a week or Moshe's ultimate challenge of limiting the work of a lifetime.

The Rabbis have been showing us the way all along—the way through the gate of limits. Through this gate of limits, we can nurture the power of voluntary simplicity and consciously reverse the trends of non-stop industrial growth. Inspired by the power of our own voices saying 'Enough!', buoyed by the Holy One, we can look into our lives and talk with our families, our synagogues and our communities and find new ways to simplify, to reduce consumption, and to advocate for the holy sacredness of our world.

As Rabbi Heschel teaches, this holy work of limit setting does not have to be the work of self-deprivation; instead, it can be the work of growing in dignity.

This Elul, how will your powers of limit making deepen?

T'shuvah and Eden

Robin Damsky

I am sitting in my yard as I write this, amidst the din of cicadas singing their love songs to one another, with the wind lusciously blowing around the 93-degree day. Although hot, the garden is nevertheless my favorite place. To my left are the grapevines yielding their first crop of grapes. To my right is a series of raised beds forming a giant U: lettuces under their shade cover, two compost bins, onions, carrots and beets, turnips and daikon, broccoli and collards, peppered with kale plants in every available space, with companions of sweet alyssum to keep the aphids at bay.

There is the tomato jungle; I call it thus because picking the fruits takes me into their internal forest. Only one sunflower sprouted this year, but she is majestic. Near her is the peach tree, barely able to hold up her arms right now for the weight of her pearly orangey fruits. Surrounding her are mints— chocolate and orange, spearmint and peppermint—fennel, peas and beans and on to the baby fig tree with its first set of figs: twelve. Not bad for a first-year crop, and on a tree that's just over three feet high. Aah. I exhale at the grace that God provides through these food plants.

This is only one side of the yard. Last year, I began the project of turning my yard into an organic, edible landscape. Culinary herbs, fruit and nut trees, medicinal herbs, fruit shrubs, and plants with edible parts such as rose hips, ferns and violets abound. There are artichokes—Chicago's not their normal habitat, but I'm giving them a shot. Abundant berries, winter squash, cucumbers and watermelons. And the Jewish connection: parsley, of course, hyssop and horseradish. It is a work in progress, but it is immensely fulfilling, a personal Eden.

Our tradition tells us that we were exiled from Eden to toil the soil, but I find that working the earth brings Eden. The sounds and smells, the growth and the creatures are all a piece of Eden. As a rabbi, one of my main visions is the idea of a time when all life—human, animal and plant—can live together peaceably. The garden contains all the ingredients for that. It is a place where we learn respect, because we know the potential myriad of setbacks: weather, "pests," soil conditions, to name a few. It is a place where we learn gratitude. How happy is the child that picks her homegrown carrot? The boy who chews on the first string bean that he planted himself?

That's just the beginning. Sharing produce with the homeless is an important gift to give each season, through our local food pantry and homeless shelter. Bringing in community members who need extra income to work the earth helps those in need while teaching them about how to grow their own food. Simultaneously, we build community across all lines: religious, ethnic, racial, even across the generations. I have made some wonderful friends through the garden that I never would have met otherwise. And we also teach our children in the arenas of Judaism, nutrition, and sustainability.

What is *t'shuvah* if not return? Are we not working to return to the source—to God, to oneness, to wholeness with all of creation? The shofar rings out each day of Elul: Awake! Awake! Awaken to our work to create wholeness.

How interesting it is that the Messianic Age is the very vision of Eden. What better way, then, for us to do the work of the High Holy Days than to plant? We bring ourselves closer to God's creation and advance the healing that pulls us a little closer to the Messianic Age. Plant and tend. A little Eden awaits you.

Shanah tovah—a year of wholeness for all.

Anticipation

Judith Felsen

> Your gentle breezes sensed in Av are invitations
> whispering of meeting in Your fields
> by trails we carve, inner journeys of *t'shuvah*,
> clearing, cutting with correction, paths of prayer
> our destination, oneness, You.

Av's warming sun evokes desires
nighttime glitter, starry cairns and
shimmering nights inspire minds
deep reflection of soul's work
in study of Your Torah.

This moon wanes, anticipation grows
yearning closeness, love's connection, our rendezvous.
Grateful heels, knees bent, welcome walk to You
blessed by joys and admonitions,
love guides the way.

The darkened moon harkens meeting
Aliyah in Your fields
grounds sown, *t'shuvah* grown,
harvest of correction, blessed by Your Forgiveness,
farming always with Your love.

Elul 4

Ready for Withering Flowers

Sarah Chandler

I'm familiar with your story
This gratitude you cultivate helps ground you
And yet, do you really deserve to ask for more?
The answer to this question will give you the balance you seek.

Sometimes you need a reminder that we already said farewell to the month of Av
As it is written: "Man born of woman is short of days and fed with trouble. He blossoms like a flower and withers, and vanishes, like a shadow." (Job 14:1–2)

In Elul, you are instructed to enjoy the ephemeral beauty of the flowers without worry of their withering
Since *t'shuvah* (repentance) is the name of the game, instead of fearing change we welcome it in.

Every morning the shofar calls you to *t'shuvah* (repentance)
Are you listening?
How might you be more awake in order to hear its sound?
Allow the August blossoms a chance to bring you to the presence you desire.

Step 1: Gather flower petals into a large bowl—ideally four colors and four different species. Bowl is ideally wood but can also be glass or metal. In New England this is a great time of year to find a diversity of lilies, Queen Anne's lace, chicory and aster.

Step 2: Fill your bowl with water covering the petals—ideally spring water but tap water is also fine. The chance to visit a river, lake or small spring will only add to the ritual.

Step 3: ASK FOR SOMETHING. This is for real. If you're going to open up enough to do real *t'shuvah* (repentance) this year, you have to acknowledge that you are not yet whole—that there is something about yourself you want to change, or at least cultivate. A useful formula is "May I be…" or "Let me be…"

Step 4: Pour the entire bowl of petals and water over your head and proclaim: "*Horeini Ya Darkecha*, הוֹרֵנִי ה' דַּרְכֶּךָ, Reveal to me Your path." (Psalm 27:11) This is both the sealing of our request and also a letting go of wanting only one thing.

Based on the teachings of the Eish Kodesh, Rabbi Kalonymus Kalman Shapira.

Personal Reflections for Elul: Mindful Turning to the Path of Love

Jeff Foust

For me, the key to the entire Jewish New Year period comes in the month of Elul, the Hebrew month which precedes Rosh HaShanah. I often have heard people complaining about being weighed down by all the emphasis during the holiday on the mistakes and wrong doings in our lives that we need to repair. Elul reminds us that the very foundation for the spiritual work that we do at this time of year is returning (in Hebrew *t'shuvah*) to a loving relationship with the Source of all life, with each other, with our own higher selves, and with the ecology of our Earth household. Feeling this love provides me with both the support and the motivation to reciprocate the love and to work through and resolve anything that's getting in the way of that loving relationship. Then even if someone else is still unable to open their hearts, at least my heart can be open.

This emphasis on love is at the very center of the meaning of Elul, which comes from its initials which point to a passage from the Biblical Book, the Song of Songs 6:3: *Ani l'dodi v'dodi Li* (I am for my beloved and my beloved is for me.)

Becoming aligned and feeling at one in love is also a central part of our ritual of blowing the shofar, which is done every weekday morning in Elul. We begin and end with the call of *Tekiah*, the single blast of the shofar, which powerfully resonates with our underlying oneness in love, and with the reality of love and reconciliation always being there for us if we only open our hearts to it. The other two broken blasts of the shofar, the *Shvarim* and the *Teruah*, represent the broken places in our lives, and are bookended and held in love, while with the support of the One Who is the Source of all life, love, and blessing, we work on our *t'shuvah*, our return to living in loving relationship.

Elul 5

Gratitude

Katy Z. Allen

Gratitude. Gratitude was where an amazing woman I met as a hospital chaplain had landed three days after learning her cancer had spread throughout her body and nothing more could be done. Gratitude—not fleeting, as she had experienced it previously—but minute-by-minute gratitude for all the goodness in her life. Gratitude with full awareness of all the difficulties.

I had gone for an early morning walk. When I reached the open meadow on the trail to the river, and the dew on the spider webs caught my eye, I felt it—gratitude. I thought of the words Jewish tradition teaches us to say first thing in the morning: *Modah/eh ani lifanechah* (I give thanks before You…) What a powerful way to begin the day!

Jewish tradition also teaches us to say 100 blessings a day, to praise God 100 times every day. Through recitation of the thrice daily prayer services, blessings before and after eating and after using the bathroom, one can easily accomplish this goal. But what if, instead, we were to say a blessing at regular intervals throughout the day. In 24 hours, that's a blessing about every 15 minutes. But most of us are not awake 24 hours a day (thankfully!). Doing a little more math, if I sleep 8 hours (for which I am grateful), that leaves 16 hours, or 960 minutes. $960 \div 100 = 9.6$. If I say a blessing every 9.6 minutes during 16 waking hours, I will say 100 blessings a day.

That is almost constant praise and blessing for the gifts bestowed by G!d—close to constant gratitude.

What would it take for me to come close enough to G!d to feel constant gratitude? Must I be dying? I certainly hope not!

May this month of Elul, of closing in on the *Yamim Noraim*, the Days of Awe, bring each of us closer to a constant sense of gratitude.

Elul: The Month for Climate Action

David Krantz

Tekiah! In Elul, we hear the call for the quintessential sound of the shofar every morning. It's meant as a daily wake-up call to action. Perhaps appropriately, the word *Tekiah* itself also means "disaster." Day after day in Elul, the shofar shouts: "Disaster! Act now!"

Just as an alarm clock gives us notice that we have to get to work, the shofar reminds us that time marches onward and that our mistakes won't correct themselves. We must actively engage with the world to repair it and our relationships with each other. The process of repentance and repair starts with recognition, and it's time that we recognize that with human-induced climate change threatening the Earth as we know it. Our relationship with our environment is greatly in need of repair in order to avert disaster. But how can we repent and repair our relationship with the Earth? Every day of Elul, we can take one step forward to mitigate and abate climate change.

You can start small, with reducing your energy consumption at home; walking, biking and taking more public transportation instead of private cars; and most simply and effectively by simply reducing your consumption of meat. Yes, that's right, eating less meat may very well be one of the most impactful ways that you can reduce carbon emissions, since meat production and consumption—more than transportation or home-energy use—is, according to the United Nations, the single largest contributor of greenhouse gases to the atmosphere. Collectively, changing our diet can help change the climate.

And think bigger! Next election, vote green by supporting the politicians with strong climate policies—and hold them accountable if they take office. Call your elected representatives and tell them to take action on climate change. You can magnify your impact by joining with others and becoming more involved with the Jewish-environmental movement, including organizations such as the one I run, Aytzim: Ecological Judaism.

People say, "Think global, act local," but that may not be enough in the era of the Anthropocene, the time when we humans have become the greatest force on the Earth's systems. We need to both think and act locally, nationally and globally. It is time for us to approach wicked problems such as climate change with multiple solutions that work across multiple levels. Each of us needs to work both within our own communities as well as in cooperation with others.

As Jews, we have a religious obligation to serve and guard the Earth (Genesis 2:15)—a responsibility we have neglected for too long. Elul is the designated time in our calendar for us to repent for our sins, but repentance in Judaism includes more than mere recognition or apology: Repentance also means being confronted with the opportunity to make the same mistake again and choosing differently. In Judaism, repentance means behavior change. And when it comes to upholding our responsibility to serve as stewards of the Earth, stewards of God's Creation, we are given a new opportunity to choose more wisely with the dawn of every new day. But unlike when we as individuals sin against our friends, our repentance with the Earth is societal, and our success is dependent on collective action.

Each of us needs to act in concert. So, both change your individual behavior and spread the word! Start climate-action conversations with your friends and relatives, and discuss climate action at synagogues, JCCs and schools. Listen to the imperative of the shofar's daily blast: "Act now!" "Act now!" "Act now!" And heed the shofar's call to action to avoid disaster. Elul is the month for our repentance, and as such it is, more than any other month, also the month of climate action. *Tekiah!*

Three Levels of Holiness

David Seidenberg

In the Torah, three things are called *shabbat shabbaton*—the seventh day, Yom Kippur, and *Shmitah* (the Sabbatical year).

Agnon, in his book *The Days of Awe*,[1] shares a teaching from Rabbi Tzvi Hakohen of Rymanov about this. The rabbi was asked, if both Yom Kippur and the Sabbath itself are called *shabbat shabbaton*, how is Yom Kippur more special? And he answered, the seventh day is called *shabbat shabbaton l'adonai* (a sabbath of sabbaths for God). Yom Kippur is called *shabbat shabbaton lakhem* (a sabbath of sabbaths for all of you). On Yom Kippur, we don't just reach toward the divine realm, we draw it into ourselves.

When Rabbi Michael Bernstein shared this teaching with me, he added: "By that logic, *Shmitah*, which is called *shabbat shabbaton la'aretz*, a sabbath of sabbaths for the land (Leviticus 25:4), draws that holiness into the land. In this way, *Shmitah* is even more akin to Yom Kippur than it is to Shabbat."

There's a midrash that can explain this idea. The essence of the *Shekhinah*, the Divine Presence, was originally in the land, in the Earth. When Adam and Eve ate the fruit, breaking God's command and sinning against the tree, the *Shekhinah* fled away from the Earth to the first heaven. With each successive generation, the *Shekhinah* fled further, until she was seven heavens away from the Earth. Then Abraham and Sarah came and drew her down to the sixth heaven, and Isaac and Rebekah drew her even closer, to the fifth heaven, each successive generation bringing the *Shekhinah* down, until Moses finally brought her "from above to below." (Genesis Rabbah 19:7)

---

[1] Agnon, Shmuel Y. *Days of Awe: A treasury of Jewish wisdom for reflection, repentance, and renewal on the High Holy Days.* Schocken Books, 1995.

But Yosef Gikatilla, the 13th century Spanish Kabbalist, explained that this didn't complete the process: "Moshe our teacher came and all with him and they made the *mishkan* (Tabernacle) and its vessels. And they repaired the ruined channels, and…they drew living water. And they made the *Shekhinah* return to dwell/*l'shakhen* among the creatures below, in the tent—but not in the ground/*baqarqa*, not in the Earth itself, as she was in the beginning of the Creation."[1] [2]

This is what it means when God says to Moses, "Make me a sanctuary *v'shakhanti b'tokham* (and I will dwell among/within them)" (Exodus 25:8): God said that the *Shekhinah* would "dwell in them," but not (yet) in the Earth. There was one more step to go.

The *Shmitah* year, when we are commanded to rest the land and to rest along with the land, when we share food and land not only with the poor and the stranger but also with the wild animals, bridges that last step. *Shmitah* is a *shabbat shabbaton "la'aretz,"* not just "*lakhem.*"

Shmitah infuses *Shekhinah* into the Earth itself. Of course, the Earth is already filled with *Shekhinah*. If we have inured ourselves to that, *Shmitah* can open our hearts. But first we need to make *Shekhinah* dwell within us, so that our hearts can meet the world "*ba'asher hu sham,*" at the level of holiness that is already there. That's what Yom Kippur does.

Sabbath, Yom Kippur, and *Shmitah* represent progressive stages of bringing *kedushah* (holiness) and *Shekhinah* into this world, from God, to us, to the Earth itself. May this be our life's work.

---

[1] Gikatilla, Yosef. *Sha`arey Orah*. Warsaw: Argelbrand, 1883.
[2] Gikatilla, Yosef. Gates of Light: *Sha'are Orah*, Translated by Avi Weinstein, HarperCollins, 1994.

Elul 6

Living with Change

Howard A. Cohen

"The end of the human race will be that it will eventually die of civilization."—Ralph Waldo Emerson, 1870[1]

With the approach of the season of *t'shuvah* it is once again time to reflect on our relationship with the Earth. In the past I asked myself questions such as: 'Did I waste natural resources?'; 'Was my carbon footprint unnecessarily large?'; or 'Did I speak out against corporate environmental abuse loudly enough?' But now I believe there is another set of questions even more important. Two of these questions are 'How prepared am I to live in an ecologically changed/damaged world?' and 'How am I helping others cope with the environmental changes that are now a part of our reality?'

The overwhelming evidence is that we have already inflicted irreversible damage to the ecology and environment of the Earth. Perhaps we can mitigate future damage a little, but we cannot undo what has already been done. This is why I think the most important existential challenge today is learning how to live in the rapidly changing environment of our world.

Sadly, the environmental movement of the last 50 years has only had limited success. This is not because truth and science are not on its side, nor because it lacked resources or organization. As a messianic movement it focused on final outcomes: If we don't change (*t'shuvah*) our ways, terrible things await us (think Jonah and his commission from God to the Ninevites). But if we change (*t'shuvah*), we can avoid a horrible fate and enjoy heaven on Earth, a return to a time when the Earth was much more like the days of the Garden of Eden. (Think Shabbat as a taste of the *Olam HaBah*, that is, being in the Garden of Eden.) The environmental movement failed because it was essentially a messianic movement and ultimately *all* messianic movements fail.

I admit that dark messages like this are not what people like to hear. Yet, if we prepare for the changes scientists are quite confident will almost certainly come, then our future need not be dire. That is why this year when I reflect on my Earth/nature relationship, I am going to ask how we can live in an ecologically and environmentally changing world. If we are not afraid of the unknown and change, then this adventure has the potential to empower, inspire hope and stimulate creativity.

---

[1] Emerson, Ralph W. *Society and Solitude: Twelve Chapters*. 1870. (Classic Reprint), London: Forgotten Books, 2012.

Fields Open

Judith Felsen

There is a sadness only we can mend
a problem only we can solve
recognizing our harvest
yielding crops of tragedy.

The foods we make we should not eat
our spoiled waters poison us
our air became a filthy haze blocking the sun
asphyxiation, suffocation is our fate
by our hands alone.

Do we see our temple walls destroyed?
Do we see the light between infested crumbled stones?
Elul we walk the grounds we hurt
amidst Earth's aching cries, delivering, loud toxic labor
undefended, pregnant, by our seeds of greed.

Our brokenness connects us with our land
fields open greet our King mistakes undone
waste compost, errors teachings, Elul of correction
sees the fox, remembers, bears hoe and hands to labor
toward our greener days, our time and King have come.

The Important Ten Percent

Judy Weiss

Rabbi Dr. Judith Hauptman, professor of Talmud at the Jewish Theological Seminary, taught a passage from the Babylonian Talmud, Shabbat 54b-55a, in a study session for the Israeli Knesset in 2014. In this passage, the rabbis conclude that we're responsible for protesting when we observe someone doing something that is morally wrong. We must protest even if we think the offenders won't heed our warnings, and even if we fear being stigmatized for speaking out.

The Talmudic passage teaches that if we fail to protest a wrong-doing that we observed, our name becomes attached to the deed because we are just as culpable as the wrong-doer. Hauptman concluded the lesson by emphasizing that to be a good Jew, it isn't enough to keep Jewish rituals and laws—one must also identify ways to fix the world and then protest until wrongs are righted. Speaking out extends beyond moaning and crying around one's dinner table. One must protest in one's neighborhood, city, to the head of state and every one of his/her aides, and throughout the whole world.

Peter Gleick, an environmental scientist specializing in energy, water and climate change, made a similar point in 2010.[1] He suggested that climate change disasters be named after climate change deniers. His logic was that deniers are stalling action to cut emissions, so our society hasn't addressed climate change adequately, and the probability of extreme weather events has increased. By naming climate disasters after deniers, we blame those responsible for increasing the odds of these catastrophes.

Yet, the sad fact is that it's our fault, and all our names should be on the disasters. If we had protested that Congress was listening to fake scientists instead of heeding the warnings of real climate scientists, then Congress would have enacted legislation long ago. If we had protested and thus created a support system for our nation's climate scientists, they would not have had to endure abuse at the hands of misleading, badgering, disrespectful, and wrong (yes, sinful) Senators and Representatives. Our use of energy and resources would have been fixed, modernized and de-carbonized years ago. We could have started working to cut emissions more effectively back in 1988.

The Yom Kippur *Al Het* prayer, written in the plural, reminds us that we are responsible for forming an ethical and just society (See the morning haftarah, Isaiah 57:14-58:14). Summarizing from the Silverman *Mahzor*, the prayer says we sinned by compulsion or by our own will, we sinned unknowingly or knowingly, with speech or hardened hearts, by wronging neighbors, by association with impurity, by denying, scoffing and by breach of trust.

What greater breach of trust could we do to present and future generations than by pushing the climate past tipping points? We sin when: we pretend we have no choice, the problem is too big, we're afraid to speak about it, with stiff-necks and confused minds we allow impure air and water to continue to hurt people...we still deny, delay, dis, digress...and break faith as a community.

Social scientists have found that when "just 10 percent of the population holds an unshakable belief, their belief will always be adopted by the majority

---

[1] Gleick, Peter. "Global weirding: Naming climate change disasters after the deniers." 17 Aug. 2010, thinkprogress.org/global-weirding-naming-climate-change-disasters-after-the-deniers-e0ca2f9a4017/

of the society."[1] For change to happen, 10% of the population must be "committed opinion-holders." So speak about climate change. Go on marches. Write to newspapers. Protest in your Senators' and Representatives' offices.

Ten percent doesn't sound like so much. But if you aren't vocal and committed, then we won't reach the 10% tipping point, and Congress won't act. Imagine if one day your grandchild asks: What did you do after each of the deadly floods, fires, and blizzards to prevent more climate catastrophes? Will your name become "mud"?

We are all in this together. But together, we can get out of it.

The time to protest is now.

---

[1] "Minority Rules: Scientists Discover Tipping Point for the Spread of Ideas" 26 July 2011, scnarc.rpi.edu/drupal6/content/minority-rules-scientists-discover-tipping-point-spread-ideas

Elul 7

Journey to a Mountain Pond

Katy Z. Allen

The word *makom* in Hebrew means "place," or "space," but it has also come to be a name of G!d.

Some places take on more significance in our lives than others. They touch us more deeply or are associated with significant memories. For me, one of these is a place I have come close to, but have not yet seen with my own eyes. Yet, just through proximity, it has touched me deeply, shifting something in my soul.

The name of the place is Gamawakoosh, but you cannot find it on a map. Gamawakoosh is the name given to this place by my mother's family.

Beginning in the early 1920s and from a very young age, my mother, her older brother, their parents, friends of varied ages, their dogs, and their nanny goat hiked for three days, with the men and boys doubling back for a second load, up the side of a mountain in the Adirondacks to a hidden pond. There, with permission from the landowner (at that time it was not public land), they built a small log cabin. They carried in all their provisions, including tools and rolls of roofing—one year the collective weight of the packs was 512 pounds.

Several journals of trips to Gamawakoosh remain intact, providing clues to the travelers' route and insight into their experiences. Stored in my memory are the stories my mother told of Gamawakoosh, her most favorite place in all the world (and she travelled to many lands during her childhood and youth). For her, it was a magical place of sheer delight, of good fellowship and long conversations, and of the wonders and awe of the wilderness. It was a place of healing and joy. August and Gamawakoosh provided a refuge from the father who at home in "civilization" was the source of emotional and spiritual pain that my mother carried with her all her life; for in the wilderness, away from societal norms, he was a different person, one she could respect, appreciate, and enjoy. Even at age 90, her eyes still twinkled when she spoke of Gamawakoosh, and it remained a place of respite for her mind and soul when her body no longer permitted her to explore the woods and fields in the way her spirit needed.

One summer a few years after my mother passed on, together with one of my brothers, a cousin and her husband, one of my sons and his wife and their dog, two descendants of another youthful 1933 Gamawakoosh participant, and a gem of a hiker who had been that man's good friend for many years, we went in search of this hidden spot. Although we tried, circumstances prevented us from reaching the actual site of the cabin, but in the process, we walked where our families and their friends had walked, and we waded streams they had forded. Although we never laid eyes on Gamawakoosh, we touched its essence. We found it in the woods and beside the river. We found it in the colorful mushrooms of the damp forest and in the fairyland nooks and crannies of mosses, ferns, and tiny pine saplings. We found it in the decaying 1939 Chevy we stumbled upon, mysteriously abandoned far from any current road. We found it in our shared breakfasts, lunches, and dinners, and in the preparation and clean up. We found it in the laughter and camaraderie that flowed among people who had never before met, and in the stories of family members long gone, whose spirits hovered among us. We found it in new definitions of family, in healing long-held sadness, and in new-found joy. And now we find it in our shared memories of a sacred place as yet unseen by our eyes.

HaMakom—The Place. The gift, the sacredness of Gamawakoosh is not inherent, but flows forth from what we do with it and what we make of it, and in the Presence that fills all space. May we all find places that become for us Places that bring healing, laughter, and new depths of love and relationship with those we know and with those we don't know. As we journey through Elul, may our hearts and souls re-turn to The Place, *HaMakom*, and to the spaces It fills.

The Known and the Unknown

Anne Heath

I celebrated my first Hanukkah amongst my siblings and their children celebrating yet another family Christmas. We had gathered for winter break in Santa Fe, New Mexico, at our brother's home, glad to be together after travels of varying distances and difficulties.

My lengthy, made-it-in-one-day drive from St. Louis culminated in a wondrous night sky display. My younger daughter and I approached Santa Fe well after midnight. The cold, crisply clear night made for perfect night-sky viewing, too good to be just an out-of-the-window-on-our-way-somewhere experience.

I stopped the car. We got out, glad to be standing. We stretched our road-weary limbs, all the while looking up in awe. We both agreed that it almost felt as if the sky were falling because the sky was so full of constellations and planets. The area's elevation made everything seem just that much closer.

Upon awakening late the next morning, we discovered that the bright, clear sky of the night before had been replaced by low-hanging gray clouds and occasional fog. Disappointing? Yes, but not nearly as problematic as what I perceived as "wrong" with the area's trees, grass and dirt/soil. The pinion pines were short and stubby. There wasn't much grass—green or otherwise. The dirt/soil was sandy clay. Nothing like the tall trees in St. Louis, nor the prevalence of lawns and dark, rich soil there. Nothing like the wide variation in flora in St. Louis—even in winter.

The brilliant night skyscape seemed "just right" immediately. The "wrongness" of the Santa Fe landscape didn't turn into "maybe this is OK" until almost the end of our visit, eleven days later.

I wondered on the drive home if my feeling of no longer fitting in at family holiday celebrations might have colored my feeling of not feeling at home in the Santa Fe physical environment.

I continue to wonder why I can get so stuck in needing my environment to be one that is comfortable and familiar. The push and pull between the lure of the new and the ho-hum-ness of the everyday is a recurring theme in my life.

In the lead up to Rosh Hashanah and Yom Kippur this year, it will be worthwhile for me to revisit the question of balance between the security of the known and the insecurity of the unknown, especially when the unknown represents new growth, renewal and health for my relationship with myself, with G-d, and with others; and even more especially when a trek off into the unknown represents a running away from what's difficult in the midst of the known—something which needs healing.

If this is your experience, I pray that the coming year will be a year in which a clarity as brilliant as the cold winter night sky outside Santa Fe illuminates your path.

Elul 8

Return to the Land of Your Soul

Adina Allen

"To serve it and to guard it." (Genesis 2:15) This, we are told in Genesis, is human beings' purpose in the Garden of Eden. Though this seems perhaps a straightforward task, in the rabbinic imagination there are many possibilities for what, exactly, God intended for humans to do in this role. One explanation put forth is that the first human was given the practical task of keeping the garden alive and healthy. In this view, human beings were meant to be caretakers, watering regularly so that the plants would grow and perhaps protecting the vegetation of the garden by keeping the animals out. However, I think there may be another, more thrilling, motive for why caring for the garden is the task first given to human beings—one that the ancient rabbis did not explore.

Through the physical act of gardening, we are not only tending the land, but we are tending souls. There is an intrinsic relationship between cultivating the soil and cultivating the self. As we work on transforming the Earth on behalf of plants, we are, ourselves, transformed.

Anyone who has had the privilege of tending a garden through all the seasons knows the magic that can be found in this enterprise. Being connected to a piece of land over a period of time provides one with constant opportunities for noticing. We become attuned not only to big, beautiful changes like bursts of colors when the perennials pop up for the first time, but also to the subtle day-to-day or even hour-to-hour changes of seedlings sprouting, working their way up through the soil, unfurling tender green leaves and pulsing down grounding white roots.

Tending a garden also helps us to cultivate patience. We learn to wait, to watch, and to appreciate the growth that happens in its own natural time. Over the cycle of the year, the garden reveals to us the truth that change and growth are constantly happening. Even in the bitter cold of winter, covered by mounds of ice and snow, garlic planted in the fall takes root and flourishes underground, hidden from our watching eyes. In the spring, we see the bright stalks of green shooting out from the soil, but the seeds are alive and growing before these signs of life became visible on the surface. As gardeners we have the opportunity to be constantly inspired by the strength and humbled by the fragility of new life.

May we answer God's call to tend to the Earth on which we stand, and the new life—both real and metaphoric—that unfurls. May the tools for growth and meaning-making that Judaism offers in such abundance—prayer, study, celebration and service—be the tools we use to tend the garden within. May we clear away the weeds that no longer serve, cultivate patience as the seeds within us germinate, and, in so doing, cause the garden of our soul to flourish.

For Lifts

Nyanna Susan Tobin

Wherever I sit or stand, it is sacred ground. Sometimes it is hard to believe this wisdom. But, if I can re-remember my roots and my strong belief that we are all a part of the on-going cycles of creation and of unraveling, then I can wake up and realize the miracle of this moment.

One of my goals for this summer was to slow down and honor my desire for living closer to the land and water in my neighborhood. But in between watering and harvesting for a few backyards, I traveled all over New England. I went to a Slow Living Summit in Brattleboro, Vermont. I found and sold an antique (1740's) map in Hyannis, Massachusetts. I went to a Social Justice and Storytelling gathering in New Hampshire, and I experienced the strong Earth energy at The Round House in Colrain, Massachusetts. We also celebrated the crossing over of two special elders who live with me in Wayland Housing. And I participated in Ma'yan Tikvah's Shabbat in Nature retreat in mid-August. As Molly B. wrote, how does all this cook nutritious bread? This summer, I have experienced so many others working to make our world work for the present and for the future.

Maybe all religions foster a love and awe of the past, our roots. The exhibit of the Dead Sea Scrolls at the Boston Museum of Science was a journey into the past. Two thousand years ago, our ancestors made pots, and ink, and parchment. They lived in uncertain times, but they left us with their seeds, the work of their hands, and their written instructions for meeting the end. The end of existence is what the scroll scribes predicted. They were getting ready for their last breaths.

What will we be getting ready for this year? We have the science, the evidence of global warming and the rise of the waters. We have people living under tyranny and those trying to take back their humanity. Our people are recovering from nearly losing our footing on Earth. I imagine that while I continue to struggle with budgets and making real food, hopefully, I will be thankful for these daily struggles and the awareness that I am standing or sitting on Sacred Ground.

Elul 9

Invitation

Judith Felsen

> My Lord, we saw Your waves
>   and thought
> You were enticing us.
> We heard Your winds
>   and talked of
> You reminding us.
> We felt Your rains
>   and whispered
> You're inspiring us.
> We fled Your floods
>   and dreaded
> You now chiding us.
> Yet all the while,
>   with Love You were
> inviting us.

Caring for the Planet

Laurie Gold

"When God created the first human beings, God led them around the Garden of Eden and said: 'Look at my works! See how beautiful they are; how excellent! For your sake I created them all. See to it that you do not spoil and destroy My world; for if you do, there will be no one else to repair it.' " (Midrash Kohelet Rabbah, 1 on Ecclesiastes 7:13)

I read this midrashic story only recently, decades after I was a teenager sitting in the pews at Temple Beth El of Great Neck, listening to Rabbi Jerome K. Davidson deliver his sermon. He was speaking about how it is against Jewish values to litter and to pollute the air and seas. He was disheartened whenever he saw people dumping the contents of their ashtrays onto the roads and sidewalks. Perhaps Rabbi Davidson was thinking about this Midrash when he said that Jewish ethics require that we do our part to take care of the Earth. Maybe he was thinking of another Jewish textual basis of our obligation to care for the environment, for we find such bases in the Bible, Talmud, Midrash and Law Codes.

Jewish thinkers in every generation and in every part of the world have urged us to care for our planet. Why, then, don't we remember? Why don't we listen? There are many reasons. Some of us fall short of our obligation because we are forgetful, greedy or ignorant. Some of us miss the mark because we are lazy, oblivious or selfish.

Fortunately, at this time of year we are given the opportunity to ask ourselves tough questions, such as: How has my conduct caused damage to the planet? How can I change my behavior so that I stop hurting, and start healing, the Earth? Can I encourage other people to make these changes too?

Changing our behavior isn't always easy. It takes time to undo bad habits and replace them with new ways of doing things. All of us have been successful in modifying our behavior in the past. We can do it again. We can make the changes needed to help improve our world. May we start today.

Growing *T'shuvah*

Maxine Lyons

I am often looking for ways to connect to *t'shuvah* even during the leisurely days of summer. *T'shuvah* for me means turning to those thoughts and actions that help me to become more of "my higher self," following those practices that nourish my growth to know *shalom* (peace) and to reach greater *sh'lemut* (wholeness). As I more mindfully pursue personal growth, I resonate to the Hebrew word, *hitpatchut*, growth through an openness and receptivity to change. In summertime, I focus on ways to practice with greater compassion in how I spend my time and focus my energy as I take on these goals.

I resonate deeply with a spiritual writing that described the personal journey of a young man who made meaningful contributions to help alleviate suffering, first locally and then he volunteered with a health organization performing basic life-saving measures for the most needy. He realized that he could not SAVE them all, that whatever he does is a small amount given the needs and intensity of the impoverishment and sickness of those in dire circumstances. And his conclusion is similar to mine—that one cannot effect major changes, but we can become more aware that individuals in pain require compassionate responses. He called it a "ministry of silence"— listening, being there, being present. I was motivated anew and started to participate in healing services for homeless people in my community in order to be a witness to their lives, to affirm their small steps to healing, to be present as they were receiving some comfort and momentary relief during the service in which I participated. One homeless woman said to me, "It mattered to me that you were here." With that comment, I committed myself to be there regularly.

My involvement with a Jewish inmate (writing him for 12 years during his incarceration) meant helping him in a variety of ways including his attempt but ultimate failure at re-entry into society after years of extreme deprivation. This included daily indignities, incivility and basic human concern. Consequently, he lacked the life skills that would enable him to succeed "on the outside." Listening to him and his travails and providing some financial assistance have given him support but insufficient for him to acclimate to life on the outside. He said, "I was physically out of prison but my mind was still shackled from the abuses."

Although there are few Albert Schweitzers and Paul Farmers capable of performing their amazingly impactful service to humanity, there are endless opportunities to alleviate the hopelessness and abject suffering of individuals in our midst. I am learning that we can offer heartfelt caring, express joy when good things happen to them, advice and empathy when the challenges cannot be faced alone and also, comforting words and deeds when they cannot succeed.

Participating in a weekly Buddhist meditation group adds to my sense of *t'shuvah*, as it prepares me to practice deep listening, offering new ways to respond with compassion and kindness and caring by being mindfully present. My deep-seated Jewish values and traditions inform how I personally address the pressing societal ills and the all too elusive peace as I learn again and again to be present a little more each year.

Elul 10

The Earth Is Crying Out in Pain

Katy Z. Allen

"The Earth is the Lord's, and the fullness thereof." (Psalm 24:1)

The Earth is crying out in pain. Yet, the beauty and mystery of the natural world shine forth, ever ready to calm us, inspire us, strengthen us, and remind us of our smallness in Creation. We walk in the woods and find wonder in the spring wildflowers. Eagerly we bite into the delicious bounty of the late summer harvest. In awe, we gain inspiration from the night sky, a sudden and unexpected rainbow, a brilliant sunset.

The Earth is crying out in pain. Yet, we climb in our cars and drive to the mall, spewing noxious chemicals into the air as we go. We buy what we need and what we want, gobbling up the Earth's limited resources, entering eagerly or reluctantly into our consumer culture that tells us that this object will make us happier. We turn on the heat in the winter and the air conditioning in summer, needing, wanting to be comfortable.

The Earth is crying out in pain. Day after day, images flash across our TV and computer screens of floods and fires and famine and drought and war. We hear catastrophic predictions of the impact of climate change on our planet. Consciously or not, fear grips us. We wish it wouldn't be so. We feel helpless.

The Rule of Context and its subset the Broken Windows Theory suggest that our microenvironment—the immediate context in which we find ourselves and the peer group in which we stand—may influence how we behave. The Bystander Syndrome predicts that if we collapse of a heart attack in a public place, we are more likely to get help if just one person is nearby than if there are one hundred. Most will walk by us. After one person has stopped to help, only then are others also likely to stop.

The Earth is crying out in pain. How can we fix the Earth's "broken windows" and fill its "abandoned buildings" so that we stop committing crimes against it? How do we find the courage to be the first to stop and help the fallen stranger, our planet?

The Earth is crying out in pain. As we engage in *t'shuvah,* as we re-turn, as we turn again and again and again toward all that is holy in life, let us hear the Earth's cry and not be afraid. Let us band together with our neighbors to transition to a more resilient and gentler society. Let us find the courage and the strength to stop rushing and to extend a helping hand to the broken Earth. Let us remember that even if we are not guilty, we are responsible. Let us take heart with the knowledge that every journey begins with the first step. Let us know in our hearts that we are not alone.

"The Earth is the Lord's, and the fullness thereof" (Psalm 24:1)

Earth Rituals

Molly Bajgot

> "This is what rituals are for. We do spiritual ceremonies as human beings in order to create a safe resting place for our most complicated feelings of joy or trauma, so that we don't have to haul those feelings around with us forever, weighing us down. We all need such places of ritual safekeeping. And I do believe that if your culture or tradition doesn't have the specific ritual you are craving, then you are absolutely permitted to make up a ceremony of your own devising, fixing your own broken-down emotional systems with all the do-it-yourself resourcefulness of a generous plumber/poet."
> —Elizabeth Gilbert, *Eat, Pray, Love*[1]

This quote came into my life by way of the TV screen. Elizabeth's words resonated with me and my own spiritual practice: looking to Judaism as the structure for my spiritual roadmap, and rounding out my ritual needs by creating Earth-based ceremony.

This time of year, our Jewish tradition offers some incredible ritual structure and technology to help us reflect and take stock of the past year. It's a time to reflect on ourselves in an effort to return, to ourselves and to G!d. A time when—as Elizabeth puts it—we have a ritual container for "our most complicated feelings of joy or trauma." To pause and recycle, much like the cycles of Earth.

One of the best examples we have of cycling and return is the Earth. She freezes and thaws; heats and cools; water pours down and evaporates up to come down again. She builds and births and then decays and renews. She keeps going, always finding resource and creating anew.

In the practice of *t'shuvah*, returning, I have looked to the Earth to support the challenging process of letting things go, recycling, and renewal. My experience of finding wholeness has been challenging. Even after the apologizing to people I have harmed, the guilt, missteps, and misgivings still tend to haunt me, shrinking around in the corner of my mind long after I

---

[1] Gilbert, Elizabeth. *Eat, Pray, Love*. Penguin, 2006.

attempt to ritually dismiss them. I longed for a way to really set things down, some place to hold the shrinking thoughts until they could decay and come back as fresh life.

From that place, this ritual came to me.

I started by walking in the woods, practicing *hitbodedeut*, returning to an outdoor sit-spot and speaking out loud to G!d. Then I noticed something: *what if G!d is in the ground?* What if I can do *t'shuvah*, complete my process and gain more palpable wholeness, by connecting to the Earth? What if *she* could help hold things with me? I began wondering if I could hook up to Earth's circuit and give away some of what I am carrying so that I can be cleared to do my best work.

So I tried it. And I liked it. And later, I took it to my Jewish Organizing Institute and Network Fellowship class (2014-2015). It was the month of Elul. We were discussing the King being in the field, and coming closer to our selves again as we started the new year. Sitting on the ground, I asked my cohort-mates to plant their hands in the Earth and ask her to ritually safeguard our heavy things: to take the stuck things, those that needed expansion but weren't moving by themselves. Ask her to please receive them and to work her powerful magic on them, to allow them to ride her cyclical magic, and in time, return them back to us after having undergone some alchemy.

That was four years ago. Still today I return to the woods to do my *t'shuvah*, to make my own process complete. We can do this ritual to connect to something much bigger than ourselves. We can do this to cultivate the ability to *rely on Earth with all our might*, and to gain *emmunah*, faith, in our ability to not do things alone. We can do this to ask ourselves the question and cultivate the curiosity, *what if G!d is in the ground?*

A Broken Sewer Pipe

Maxine Lyons

When a sewer pipe broke under our home directly affecting our front gardens and lawn, and a crew came in to excavate nine feet down to access it and repair it, fillers from the Earth's bowel began to surface—tons of rubble, debris, clay, stones and brick. I felt incredulous, how can this be happening as we were adding final touches to our lawns and gardens in our front yard and simultaneously renovating our large back yard lawn all in time for my son's wedding celebration! What bad timing at a great cost and unnecessary distraction from the important issues of a wedding.

After the initial shock and disgust with the torn-up yard and the putrid waste, I started to reflect on this milestone event. How can I regain my focus on this marriage as I leaned over on bended knees removing the runaway stones and clay pieces in my garden? I soon noticed that the ground could rejuvenate quickly once I cleared the debris, added new loam and nutrients and marveled at how forgiving the Earth could be. The reality became less serious despite the large output of money and time. I found several young men to assist me in the digging, removing debris and replanting, and saw that they were taking some delight in helping in the landscaping. Changes started to occur.

Slowly, I felt some internal changes: the recognition of the parallel spiritual regenerative energies happening so that by adding enriching soils, grass seeds, and colorful flowers, these beautifications corresponded to the inner cleansing and nourishment that I also needed to ensure my own new growth. I imagined the wedding festivities and the goals of marriage alongside the spiritual work I needed for my *t'shuvah* work for the High Holidays.

Songs began resonating in my head about returning to one's soul. With many turns of the soil came an awareness of turning to new and more lofty thoughts, those small discoveries which are the guideposts along the way for doing the more serious work of spiritual renewal. Each turn offered a new perspective as I continued to re-landscape my garden and re-seed my spiritual thoughts for the new year. *T'shuvah* for me means accepting life's many challenges as a way of "turning" toward the positive and relinquishing the negative habits and behaviors. I welcomed the turn toward a more elevated place in which I could form my prayers, giving new shape to my hopes and dreams for a more loving, caring attitude toward the Earth, toward myself and others, and toward holiness and the meaning of my son's upcoming marriage.

I was synchronizing the turning of my internal process to the Earth's turning. I could participate in this much anticipated wedding event with smiles.

Elul 11

Gardening Partners

Dorit Edut

This year I decided to take photos of my garden during each of the different seasons, and it is quite amazing to see the development of the various perennials and how the overall face of the garden alters. It is quite astonishing, too, to find flowers growing in places we never planted them—including a beautiful white hydrangea bush that seems to have come from an underground shoot far from its parent plant! But weekly, I also find certain weeds appearing, which I neither planted nor desired them to grow. All this only reminds me that as much as I may THINK that I am planning and planting this garden, it really is a masterpiece of our Creator, and that I am just a participant in this work.

As we begin the month of Elul with our thoughts turned towards the upcoming High Holy Days, we bring forth our own blooms and weeds of this year, some of which we may be surprised to find emerging in the patterns of our behavior, our speech, and our thoughts. Have we unknowingly cultivated these? Which ones do we want to encourage and support? Which ones do we want to cut back or eliminate? And as we recite the psalms in the morning prayers of Elul, especially Psalm 27, we are reassured that our relationship to HaShem is still intact, that the work of Elul, as Rabbi Michael Strassfeld puts it, is "to recapture a sense of self-worth based on being cherished by the Holy One,"[1] and that knowing this we are then ready to not only look at our gardens but also do the pruning, trimming, and replanting for the next year.

We humans, who are God's appointed stewards over this Earth, let us not forget who our real Partner is, who is keeping things going here, year after year. Let us align ourselves to do all that we can to live up to the great gifts that God has given us, and bring this consciousness of God's awesomeness to all that we do:

> You take care of the Earth and irrigate it;
> You enrich it greatly, with the channel of God full of water;
> You provide grain for men; for so do You prepare it.
> Saturating its furrows, leveling its ridges,
> You soften it with showers, You bless its growth.
> You crown the year with Your bounty;

---

[1] Strassfeld, Michael. *The Jewish Holidays: A Guide and Commentary*. Harper Resource Quill, 1985, p. 95.

fatness is distilled in Your paths;
the pasturelands distill it;
the hills are girded with joy.
The meadows are clothed with flocks,
the valleys mantled with grain;
they raise a shout, they break into song. (Psalm 65:10-14)

Changing Ourselves

Thea Iberall

Leo Tolstoy (1828-1910) said, "Everyone thinks of changing the humanity, but nobody thinks of changing himself."[1] We must wean humanity off of fossil fuels before the seas rise too high and before droughts have not just millions of people on the move, but billions searching for food, water and stable governments. What am I personally doing to change myself to help alleviate the problem? I drive a hybrid car and try to use less electricity. How much of a difference will it make?

In 1908, Tolstoy wrote "A Letter to a Hindu,"[2] in which he argued that it would be through love that the Indian people could become free from British rule. Mohandas Gandhi (1869-1948) read this letter and was greatly influenced to adopt a nonviolent peaceful resistance for the Indian Independence movement. A few years later, Gandhi published a list of seven social sins, the results of a correspondence with a friend Frederick Lewis Donaldson. He commented, "Naturally, the friend does not want the readers to know these things merely through the intellect but to know them through the heart so as to avoid them."[3] The sins are: wealth without work, commerce without morality, worship without sacrifice, politics without principles, pleasure without conscience, knowledge without character, and science without humanity.

I ask myself, what is worship with sacrifice? Is it going to our religious institutions one day a week to pray? Is it the *karbonot* of early Judaism, to make animal sacrifices in the Temple in order to be nearer to G-d, to express gratitude, or to atone for a sin?

---

[1] Tolstoy, Leo. *Pamphlets : Translated from the Russian*. Translated by A. Maude. Christchurch: Free Age Press, 1900, p. 29.
[2] Tolstoy, Leo. *A Letter to a Hindu*, 1908. Retrieved from http://www.online-literature.com/tolstoy/2733/
[3] Gandhi, Mahatma. *The Collected Works of Mahatma Gandhi*. Vol. 33, Obscure Press, 2008, pp. 133-134.

The Talmud says, "Deeds of loving kindness are superior to charity" (*BT Sukkah 49b*). *Chesed* (loving kindness) is a virtue that contributes to *tikkun olam* (repairing the world). In Judaism, our *chesed* actions include sustaining children, the sick, strangers, mourners, and communities. But when we worship, we aren't required to do these things. No one stops me at the synagogue door and asks me to list my sacrifices. What selfless acts am I doing for humanity and other living things? If I am to be a spiritual person, my *t'shuvah* must be to worship with sacrifice by acting through my heart. More than giving money to charity, I must change myself, and in doing so, I change the world. What are you willing to do?

Resistance

Lois Rosenthal

There is resistance to the waning of the year,
These late summer days of afternoon warmth
Sun's glare softened by a chill
A bit of orange in the solar yellow.

Fall is almost upon us, Elul is here.
Time to think about wrapping up this old year
And stepping into the next.

There is resistance.

Remember last year's beginning?
The intentions, the clarity of changes to make,
The possibilities of bringing G-d into one's life
The realizations of a path to follow.

Now we have questions to answer.
What happened to those efforts?
Some were fruitful, some forgotten
Others too simplistic or too hard.
Perhaps there were some moments of contact with the Divine
And other moments so very human.

Resistance expects an inner voice
To answer with a list of shortcomings.

Yet Elul is a tender month, and guides us
To approach the new year with appreciation for the old,
To consider every mindful effort
Then build on what has been done,
To repair the weak spots
And keep building.

Elul 12

Detroit, Our Spiritual Journeys, and Coming Back to Life

Moshe Givental

Every year on Tisha B'Av, we begin a seven-week journey of preparation for Rosh HaShanah and Yom Kippur. Like most significant experiences in life, for the Jewish Holy Days to have the potential for transformation, they require preparation. At Tisha B'Av, we reflect on brokenness of our physical, ethical, and spiritual worlds. From this, the darkest of places, we move towards hope, of a world filled with love, six days later at Tu B'Av. These are a miniature version of the journey through the next month. We spend the month of Elul in *heshbon hanefesh* (our soul accounting), reflecting on our past year, righting the wrongs we can, softening our hearts enough to apologize where needed, setting new goals, and beginning again the work of rebuilding relationships with family, friends, G-d, and ourselves.

In 2015, my journey of reflection and rebuilding started in Detroit, a city ravaged by decades via an exodus of jobs from the city after World War II, then white flight and abandoned property, then riots, crime and outrage, then political mismanagement and neglect, the recession of 2008, followed by Emergency Management's systematic undermining and deconstruction of many basic vital services such as education, city pensions, and access to water for the city's poor residents. It is a reality and history devastating to those of us who face it for the first time. However, amidst all of that, what was even more powerful for me in getting to know the city, was the way that her residents were sowing seeds of hope and life. Street Art such as the Heidelberg Project is giving residents a way to express their grief and dreams, while beautifying their neighborhoods. Residents surrounded by the blight of empty lots and decrepit buildings are getting their hands dirty and learning how to grow food. Places like The Georgia Street Community Center are putting Detroit at the top of urban agriculture in the United States, a part of the city's approximately 1,300 urban gardens and farms, while building community, the local economy, and resilience in the process.

Detroit is literally coming back to life, from the inside out, while its old top-down and government-controlled structures are still crumbling. Reflecting on this transformation, I think of our prayers for renewal and growth each day and on the High Holidays! It brings me to the *Amidah*'s second prayer, referred to as *Gevurot* (God's Might). While some Reform and

Reconstructionist prayer books interpret resurrection literally, balk, and then excise it…our tradition long ago recognized the *Mekhayeh meitim* (the vivification of life) is also metaphor for something we all experience as we grow, stumble, fall, and try again. Our prayer repeats the phrase "*Mekhayeh meitim*" three times. Therefore, the sages ask, what are the three different ways in which we fall into despair, into darkness and destruction, and might be able to come back to life? Instead of giving you their answers, I challenge us all to meditate on this question in our own soul accountings. The residents of Detroit are clearly organizing, rebuilding, and bringing their city back to life! The questions for all of us are: What do we see in our life that is falling apart? What is decomposing? What kinds of seeds, creativity, and courage do we need to plant in order to come back to life?

T'shuvah and Beauty

Lois Rosenthal

The weekly Haftarah readings follow the story of the Israelites after they crossed the Jordan into the Promised Land. The writing styles vary greatly, from poetry to historical prose.

Of particular note are writings from the time of the divided kingdom. Conquests of the Northern Kingdom of Israel and the Southern Kingdom of Judah were seen by the prophets as divine punishment for failure to follow the Torah. The writings from this time leading up to Tisha B'Av are full of harsh rebukes and biting metaphors.

Once Tisha B'Av is over and the High Holidays are approaching, the tone changes. Both Torah and Haftarah readings become infused with literary beauty—the lyrical prose of Deuteronomy accompanied by the lovely poetry of late Isaiah, filled with images of nature's grandeur as a reflection of the Divine, beckoning us to look around at the world and the heavens and there find G-d.

It is in this milieu of beauty that the month of Elul arrives, with its invitation to embark on a path of *t'shuvah*. This complex process of correcting our mistakes and learning from them can be painful and intense. The beautiful biblical backdrop for Elul and *t'shuvah* may serve to cushion a difficult process, giving hope for beauty and transcendence in resolution. It certainly honors the process.

We know that the perception of beauty affects us deeply. We crave beauty, we seek it out, we spend our precious moments dwelling on that which offers it. A Dutch still-life entices us with its intricacies and balance; intense patterns on flowers are gorgeous beyond imagination. Birds' plumage dazzles us with striking elaborations. The music of synagogue prayers draws us in; we sing and the notes hum inside us. We gaze at colors of a sunset sky; we rush outside to see a rainbow.

We perceive beauty and drink spiritual nectar—tasty, nourishing, filling. Every single human being is endowed with this faculty, through whatever sense functions within them.

On the evolutionary level, there seems to be no biological utility to this capacity we have for deep appreciation of certain "results" of our five senses. Call it a gift from G-d, a blessing.

But still, nothing in biology is maintained unless it endows the species with a way to strengthen and perpetuate itself. The biologic utility of the pleasures of food, sex, and so on, are obvious. But what about the pleasures of seeing or hearing beauty in nature or in the artistic creations of humankind?

This pleasure feels like an instinctual form of love, an immediate response on a tiny scale. Suppose you come across a wild iris in the woods. The iris is existing happily in its own environment; it doesn't need you for food or water. You find it beautiful, it pleases you. You have experienced a quantum of love for this little iris. Now you care about it. A connection has been made.

A piece of music stirs us—how beautiful! It was composed by a human being, played by other human beings. We don't know them; they may look nothing like us. And yet, some of that sense of beauty, that love we felt for the music spills out onto the humans who created it. A connection has been made.

Look out over a swath of treetops. The pattern of greens and rounded shapes is so pleasing. We can't help but love the trees, plus the whole web of nature that both sustains them and relies on them. A connection has been made.

Our ability to take pleasure from the natural world and from artistic creations of humankind creates threads of connections between each of us and the myriad elements of nature.

Beauty does have biological utility. It is an antidote to narcissism and loneliness. It connects us to the web of existence in the world, causes us to care about it, love it, and do everything we can to preserve it.

Genesis was right. We are stewards of the world. We are the only species that can preserve it or cause large scale destruction of it. Look for beauty in the world, and there you will find the passion to preserve it.

Elul 13

Movement Building and the Body

Janna Diamond

I invite you to notice where you are. Bring attention to your state of being. The way you are holding yourself up. Perhaps you choose to relax the muscles in your jaw ever so slightly right now.

Did you know that movement in the body never repeats itself? Even the most subtle motion can never be replicated. Each gesture is an expression of where you are in space and time. Movement is information. Sensation is knowledge. Every breath is change. You are here.

The body is our environment. The environment is our vaster, breathing body.

Let us become fluidly adaptable beings, softening to ourselves and those around us, generating authentic expression. Naming what we see and feel. Allowing sadness, fear, and possibility to stir and rise up within us.

In the face of incredible uncertainty, let us bring attention to how the physical body literally roots us in the precarious social and ecological conditions of our time.

How are we relating to the body? How do we care for our home?

Noticing subtleties of sensations draws us into a wider scope of awareness. We begin to practice (emphasis on practice) that which we are and want to become.

When we go slowly, we can chip away at fragmentation and detachment. We can listen to and hear where we might be needed most in this time. We can reinhabit ourselves with greater clarity, purpose and power to disrupt the systems threatening our existence.

Elul readies us for turning, returning, and change. We are made to change. Change is possible and inevitable.

In the face of all that is here and all that is to come, let us turn toward and honor our body, this body of Earth. To thank it for holding us up. To allow it to express in real-time. For renewal and resiliency. For healing. For moving forward and towards wholeness.

The Emergence of Aliveness

Natan Margalit

On Rosh HaShanah, we say "*hayom harat olam*" (today is the birth of the world). But it isn't just a birthday that happened in the past. The daily morning blessings remind us that God creates the world anew every day. So this High Holiday season is a time to celebrate a process of ongoing creation.

It brings up the question: what do we even mean today when we talk about God's creation of the world? I certainly don't mean a fundamentalist idea that God is a Being in the sky who spoke five thousand, seven hundred and something years ago and created the world. By Creation I mean that there is wisdom, beauty, value and holiness that are embedded in every atom and molecule, every particle and wave that makes up the cosmos. *Kulam b'chokhmah asitah* (You made all with wisdom) (Psalm 104:24). And, this value, wisdom, holiness wasn't just planted in Creation sometime in the past. There is something about this creating that is dynamic: from simple forms more complexity arises; new patterns, values and beauty appear that wasn't there before. Creation keeps on emerging.

When I say "emerging," I mean it technically: There is a new science of Emergence that has come along mostly in the last 30 years or so. Along with Chaos theory, Complex System thinking and other related fields, Emergence helps us to re-evaluate the way that we think about the world. Emergence basically says something new can emerge when parts come together to form a whole and the whole is greater than the sum of its parts.

So, for example, a long, long time ago there were atomic elements floating around. Two hydrogen atoms got connected to an oxygen atom to form H_2O: water. Neither hydrogen nor oxygen had the properties of water such as wetness, the ability to dissolve many things or the ability to put out fires. But together hydrogen and oxygen form something that is greater than the sum of its parts.

Molecules kept on mixing and forming more complex patterns. At some point in that distant past, the molecules reached a point when they crossed an amazing threshold—out of those molecules of matter, those separate parts, emerged an entity that was alive. The wisdom embedded in Creation had brought forth something so new it was a world apart from its component parts. We still find it amazing, even miraculous, that this quality we call life emerges from mere matter.

The crazy, creative, sometimes cruel, sometimes kind process continues. Eventually, a being evolved that could use symbolic language: the ability to create our own worlds of culture; our own environment of society that seems to surround us like a bubble with assumptions, concepts, manners and customs. This being creates not only tools but technologies that can literally change the face of the Earth, change the climate; even change our own bodies.

The Torah tells us that we are created in the Image of God—and it is true—we can be as gods, to create or destroy. The Talmud follows up on this idea and says that each human being is a world. We have the ability to create or destroy that world we call a human being.

But we also have the ability to destroy our physical world here on Earth. With our amazing brains we have evolved the ability to take apart, to analyze, break down and separate all the miraculous aliveness that has taken eons to emerge. We have raised ourselves up so high in our godliness that we imagine that we are separate from the aliveness of the world. We even imagine that the world is not essentially alive, but rather is to be likened to one of our creations: a machine. If the world is a machine we can stand above it, control and predict everything about it. We have believed that we truly are gods of the world.

But, the wisdom embedded in the world won't let this falsehood endure. We are finding that it doesn't always work to stand above and break things down. We are starting to realize, hopefully not too late, that aliveness only emerges in the connections and patterns that hold us together. We are truly partners with God in Creation—but only when we have the humility and wisdom to realize that we are a part of Creation—not apart from it. We can continue to join in, even sometimes lead, the dance of emergence. We can create and witness creation of new patterns, a kinder, more just and more beautiful world. The choice is ours: to destroy ourselves and the Earth in our arrogance, or to join in the dance of Creation with humility, creativity and joy.

Elul 14

Sweet and Sour Grapes

Robin Damsky

I am in my favorite place at my favorite time: in the garden, in the morning, before the cars have started up, before the noise of lawn mowers and leaf blowers. The crickets are singing, the birds responding. The rising sun's light filters through the leaves. A beginning.

It has been a tough year in the garden. An endless winter caused a late start and temperatures have been cooler than usual. A call from critter to critter that I cannot hear lets them know there is bounty on my corner. Maybe it's because the peach tree lost its flowers in a hard spring rain, but squirrels have eaten a fair amount of my produce this year, taking a bit of a turnip and leaving the rest (yeah, I'm not surprised). Mice, too, have traversed here. I have never seen one, but my garden helpers have. Let's not forget the birds.

At the same time, the blackberries went wild. Literally. I have cut them back and dug up new plants several times. Cucumbers abound. Arugula sings its symphony. The carrots are fat and rich. I could go on. But what hits me this year is the contrast between disappointment and satisfaction; the moments of wondering why I do this at all, pitched against the incredible feeling of gratitude when I bag up four bags of produce filled with veggies, fruits and herbs, for our local food pantry. When neighbors come by, they tell me they've been feasting on the blackberries. Who wouldn't?

This is the rhythm of the Elul and High Holy Day season, the time when we take stock. How many things did not turn out the way we wanted them to this year? How many grapes did we plant that turned sour? (Most of mine have been chomped on by critters.) What do we do? Do we become depressed or disheartened? Angry? Do we give up? Or do we plant more seeds?

Perhaps we do all of the above. Perhaps we need to feel the grief and disappointment of our losses and our failures. Perhaps we need to feel the frustration. But Elul and the High Holy Day season tell us this is only part of the process. For us to fulfill the essence of this time of year demands that we somehow find a way to get to the other side. Maybe that includes a change of project, or maybe it means finding a new way in the same project.

I sometimes think that it is all the difficulties involved in growing food that inspired our Jewish ancestry to move away from its agricultural roots. This was revived, however, with the kibbutz movement in Israel's pioneer days, and is experiencing further revival all over the Jewish world today. As we demand more sustainable lifestyles and healthier, more affordable foods, we are revitalizing our synagogue and neighborhood networks to feed ourselves and the hungry around us.

Even as I write this, I observe a critter that has found her way into the grapevines. I go over to see the culprit. A squirrel. She takes her time untangling herself from the vines, climbs up the adjacent telephone pole, and when far enough away from me to rest in safety, turns. I see the bulge in her mouth. She takes out her dessert—a nice, fat purple grape, and eats it in front of me.

Not all of our plans will fruit the way we hope or plan. But this is the season to harvest the best of our works this year, and to plan and plant again, for a fuller, richer, more bountiful harvest in the year to come.

May your Elul and the year to come be rich with new ideas and renewed energy to plant and see them bear fruit.

Elements

Judith Felsen

You gave us wind
  we hid from it
You granted us rain
  we wasted it
You made us soil
  we tainted it
You gave us air
  we polluted it
You showed us fire
  we abused it.

Our response
  has been destructive
Your response
  has been corrective.
You have given us directive
  let us use it.

Wasting Food

Scott Lewis

"When you besiege a city for many days to wage war against it to capture it, you shall not destroy its trees by wielding an ax against them, for you may eat from them, but you shall not cut them down." (Deuteronomy 20:19)

The mitzvah of *bal taschit* (do not waste) helps frame Jewish environmental concerns. While most Jewish environmental activists recognize the importance of *bal taschit* for prohibiting wasting energy and polluting the Earth, we might easily overlook the commandment's important connections to food waste.

Our sages understood this link. The Rambam, for example, pointed out that the Biblical passage was not limited to wartime actions:

"And not only trees, but whoever breaks vessels, tears clothing, wrecks that which is built up, stops fountains, or wastes food in a destructive manner, transgresses the commandment of *bal taschit*…" (Maimonides, Sefer Ha-mitzvot, Positive Commandment #6).

About 30-40 percent of food that is produced is thrown away, a shocking figure in the face of worldwide suffering due to malnutrition and starvation. The scale of food waste also has global environmental implications. Clearly, we are squandering the energy that goes into the production and transportation of food that is later thrown away. But did you know that when food waste is buried in landfills, it creates significant quantities of greenhouse gases that exacerbate climate change? The authors of the book *Drawdown* enumerate the top solutions for reducing greenhouse gas emissions and list the reduction of food waste as its third most important solution, stating, "Ranked with countries, food would be the third largest emitter of greenhouse gases globally, right behind the United States and China."[1]

Much of the food waste in developed countries occurs after the food has been delivered to retailers. Stores often throw away any "ugly fruit," thinking that a bruise may make it unacceptable to consumers, and they toss food items based on the "freshness" date, even though many foods remain edible long past those dates.

Surprisingly, the amount of food that we consumers throw away may be greater than waste at the retail level. Some of us forget to use what we've purchased, bury it in the freezer, or simply purchase too much to use in the first place and it spoils.

---

[1] Hawken, Paul, Editor. *Drawdown: The Most Comprehensive Plan Ever Proposed to Reverse Global Warming.* Penguin Books, 2018, p.42.

I find myself guilty of being part of this problem. So this year, my *t'shuvah* will include reflecting on my food waste sins: the spaghetti sauce I left on the stove, the pickles languishing in the back of the fridge, and the beautiful Russian black bread that hardened into a brick before I could enjoy it. My *t'shuvah* will also include a vow to do a better job of following the *bal taschit* mitzvah, especially as it applies to food.

Elul 15

Elul Joy and Love

David Arfa

I'd like to speak about joy and love. I know that Elul is upon us; a time for relentless self-reflection, spurred on by the blasts of shofar. And yet, the rabbis in their complexity have added another dimension to Elul: love.

Remember the acronym for Elul? It's from the Song of Songs 6:3, *Ani l'dodi v'dodi li*—I am my beloved's and my beloved is mine. Reciprocal love is spiraling back and forth right here in Elul along with our lists of how we missed the mark. Isn't this worthy of attention? What might it mean?

Here's where it takes me. "Rabbi Akiva said that if all of Tanach (the five books plus all the prophets plus all the writings) is Holy, then the Song of Songs is the Holy of Holies!" (*Mishnah Yadayim* 3:5) The Song of Songs is sensuous and loving, filled with tension, desire and yearning; lovers are seeking fulfillment in gardens and fields on every page. We all know that steamy passion can easily burn and destroy, and yet, Rabbi Akiva holds this up as the archetypal place of holiness. Blessed be.

The Song of Songs reminds us on every page that loving, sensual energy is paramount in all of our relationships—with each other, the natural world, and the Source of Life. Seeking love yearns for the reward of receiving love; and then I feel fully me, fully seen, feeling even fuller than me! I am my beloved's and my beloved is mine. Is it possible that *t'shuvah* can inspire us to reclaim this loving joy in all of our life and remind us that this is our birthright?

I heard that the great psychoanalyst Milton Erickson tells a story of a mean nasty man who never smiled.[1] He became thunderstruck and lovesick with the new school teacher in town. He asked to see her formally, and she said, only if you clean up your ways and try to smile once in awhile. The goofiest grin came over his face, kindness filled his heart and he never looked back. They lived happily ever after, smiling and holding hands like young fools until the end of their days. Here, the power of love drives *t'shuvah*.

Of course, the allure of romanticism is just that fantasy of "happily ever after." Deep down, we all know the sequence: Beauty meets Beast, Beast turns into Prince through love, Beauty marries Prince, Prince turns back into

---

[1] Bradshaw, John. *Creating Love: A New Way of Understanding Our Most Important Relationships*. Random House Publishing Group, 2013, p. 178.

Beast. (In all fairness, Beauty has her transformations too!) Undoubtedly, the mean nasty man of Erickson's story slips and falls too. Perhaps this is why the linkage with Song of Songs and Elul is so very critical. Life is also Kafka, not just Disney. Here, *t'shuvah* can be our remedy, reminding us that we can actively participate in the work of turning our sense of "unlove" back into "love."

Rabbi Akiva is saying that this great love is our birthright, there is nothing to earn. Contrary to Hallmark cards, this is very different than our relationships in this world, inside our families. I am my beloved's and my beloved is mine. Our very natural relationship with the world and God itself is to love and be loved in return merely because we are alive! This love, is a much more holistic love that applies to our entire being. Forgetfulness of this birthright of joyful loving is the way of our world. Much happens every day to blur our vision. From ordinary, imperfect attachment in childhood all the way through adulthood.

T'shuvah is like clearing our vision. *T'shuvah* helps me learn the ways that I actively block this joyous knowing; the many ways that I choose judgmentalness, pickle myself in anxious worry and bewitch myself with harsh fears. *T'shuvah* shows me that by tending to my relationships with kindness and care, I can enter the apple orchard of love once again, to know love and be re-inspired to grow love and trust once again my passions, desires, and hopes for my partner, my life, my world and my God. After all, I am my beloved's and my beloved is mine.

Reflections on the Seasons of My Mourning

Leslie Rosenblatt

Almost a year ago, my husband Marty succumbed to the ravages of cancer, leaving us on a fall evening in October. Although this event was not unexpected, we had been hopeful that he'd live longer with hospice in place. We had no idea how very sick he was and how soon the end would come.

He came home from the hospital for the last time on Yom Kippur. Just days before, we had sat for a festive meal at our dining room table. Marty was anxious to teach Nina and Gideon, our then four-and-a-half-year-old twin grandchildren, about those things that distinguished the Jewish Calendar from the more usual Gregorian calendar.

"The Jewish Calendar was determined in large measure by the phases of the moon," Marty said. "Many of our holidays come with the full or new moon."

There it was, the reference to the outdoors, the place that Marty was most content. He held up the two calendars, side by side. The Gregorian Calendar had pictures of kittens and the usual names of the months, January, February, March. The Jewish Calendar marked the year 5773 and its year began with Tishrei. "Here we are at the birthday of the world, the beginning of a new year." What could four-year-olds know or understand of all this? But we listened and we celebrated the joy of being together. We blessed the candles, the wine, the round challah and dipped apples into honey so that we might have a sweet year. These acts we did together. And then, as we had done for so many years, we ate a festive meal, brisket, potato pudding, fresh vegetables, fruit.

Marty died on Wednesday, October 3 and was laid to rest on October 7, a Sunday and *erev* Sukkot. Thus, the *shivah* was terminated by a festive holiday. How would I mourn? Two, perhaps three weeks later, I decided to return to Hamlen Woods, a conservation area where we had walked when Marty was healthy, and a place to which we returned as we thought he was healing. There I traced our steps, experienced the closeness of nature, the beautiful, radiant light of fall, and began to soothe my soul. I walked almost every Wednesday, and other days as well. My walks were meditative, reflective and full of conversation. Each day, I brought a stone or picked one from my walk. I held it in my hand, caressed it, felt its sharp edges or smooth finish. It traveled with me, through the beautiful light of fall, the biting cold and snow of winter, the heat and humidity of summer and, again, as I anticipated the fall. Acorns fell from the oaks, an occasional golden leaf from the beech was there on the path, the fall would come. I kissed the stone and placed it on the marker "Given to the Sudbury Valley Trust in memory of Paul M. Hamlen— 1960." I love you, Marty, rest well.

This year was a struggle, but I am strong. I willed my endorphins to flow—those "feel-good" neurotransmitters tide me through this time. I found comfort in these familiar surroundings. I found beautiful big birds here, the pileated woodpecker and the blue heron. Frogs shrieked and turtles slipped off logs, back into the pond. I saw the beaver and mink. I found the same friendly faces of dog-walkers, grandparents with grandchildren, even friends from the temple. I took nothing for granted. All these gave me strength and confidence and hope to continue to live life with fullness and meaning.

Ode to Hamlen Woods

The day is steamy, yet the pink water lilies stretch their petals toward the sun.
The pond is still, no cat's paws cause the surface to ripple.
The heat has caused the edges to evaporate, leaving mud at its periphery.
Look carefully and you can see the bulging eyes of a green frog,
Move too quickly and with a shriek and a jump he has vanished.
The sky is clear and blue, a small cloud passes by,

A tree swallow catches its lunch—mid-air.
From a small bridge look down, there under a lily pad,
A fish takes refuge in the cool of the shade.
What a very glorious thing to be a part of this peaceful place.

An Old Problem

Jacob Siegel

Human-made climate change and the potential it has to wreak disaster on our Earth's ecosystem is new and unprecedented. Every year, extreme temperatures rise and extreme weather events become more common. These are challenges we have never faced. They require spiritual resilience and creativity.

On the other hand, this is an old problem, a human problem: an existential crisis of our own making. We, the Jewish people, know what it means to face a crisis of existence after a cataclysmic tragedy—the destruction of the Temple—because of a moral failure of society, namely *sinat chinam* (baseless hatred).

Sometimes, it feels like humans need a threat of imminent destruction to wake up and behave better. But Jewish tradition offers an easier spiritual path—a set of debates, and codes of law, all arguing about the way we ought to live our lives more sustainably.

In fact, an entire chapter of a book in the Gemara, *BT Bava Basra* Chapter 2, focuses on the responsibilities of property owners to their neighbors and to the commons. *Mishnah* 5 outlines how far away one must keep a dovecote from a town in order that the doves not eat the seedlings in the gardens, and *Mishnah* 7 notes how far away one must plant a tree in order to protect the appearance of the city.

What kind of a problem is climate change? More than an environmental problem, it is a societal and moral problem. It requires us to behave differently. There are several different categories in Jewish law that deal with our behaviors and how they relate to the commons, and they each offer spiritual insights.

For example, there is a category of deeds, those performed in my own domain that cause damage after a delay of time. Does climate change fit into this category? The Talmud considers such deeds permissible—I am doing them in my own property and not immediately damaging others. This represents roughly where our society stands right now; an understanding of climate change as causing long-term harm, but no legal obligation to act on it.

Or is climate change more of a special case, a societal imbalance demanding intervention? The *Mishnah Brurah* (242:2) in the 19th century recounts a time when fish sellers were raising their prices before Shabbat to exorbitant amounts. He ruled that in such a case, the town should impose a decree and have no one purchase fish for several weeks until prices declined again.

Or is climate change a personal moral problem? Of a sort that even if we can't find a technical prohibition against emitting too much carbon, it might be *hayav be'dinei shamayim* (liable in the heavenly courts)?

I often feel tempted see climate change as something new and unprecedented. This can contribute to a sense of fear and desperation, a panic that can sometimes lead us to reckless choices in forming our long-term strategies. As a popular saying goes: "life is very short, so we must move very slowly."

Our *mesorah* (tradition) offers deep wisdom about encouraging individuals to behave certain ways to solve communal problems; we can use that wisdom to guide us in this new, old problem of climate change.

Elul 16

Returning to Diversity

Michael M. Cohen

The opening chapters of Genesis not only include the account of the creation of the Earth but over and over tell us of the importance of diversity. All of creation is called "good," reminding us of the value of the multiplicity of the world that we live in (Genesis 1:25). The text also teaches us, by describing everything that is created before humans as "good," that all things have intrinsic value in and of themselves beyond any value that we may place on them. Once humans are created, "very good" is the adjective applied by the text (Genesis 1:31). An anthropocentric reading of the text would say this is because the world was created for our needs, and once we are in place we can do what we want with the world. A biocentric reading of the text says that "very good" only means that creation as described in the text was complete, and that we humans were the last piece of the biological puzzle.

This reading is supported by the reality that if humans were to disappear from the face of the Earth all that had been created before us would go on quite well, actually better, without our presence. However, if a stratum of the diversity of life that had been created before humans were to disappear, we, and all that had been created after it, would no longer exist. In a bit of Heavenly humor on survival of the fittest, it is actually the smallest and least physically strong species, like the butterflies, bees, and amoebas, which hold the survival of the world in place. Unlike the other species of the planet, we have the power to commit biocide if we do not protect and preserve those smaller forms of life.

The importance of diversity is emphasized a few chapters later, in the story of Noah, where Noah is told to bring pairs of each species onto the ark so that after the flood they can replenish the Earth (Genesis 6:19). After the flood, God places a rainbow in the sky as a reminder to never again destroy the world. It is both a symbol and a metaphor: a single ray of light refracted through water, the basic source of all life, produces a prism of diverse colors.

Immediately following the story of Noah we read about The Tower of Babel (Genesis 11:1-9). The whole account takes up only nine verses. The conventional reading is that its message is one against diversity; the babel of languages at the end of the story is understood as a punishment. The Israeli philosopher Yeshayahu Leibowitz presents a different reading of the text.[1] For Leibowitz, Babel represents a fascist totalitarian state where the aims of the state are valued more than the individual. In such a society, diverse thought and expression is frowned upon. The text tells us that everyone "had the same language, and the same words." (Genesis 11:1)

We read in the genealogies that link the Noah and stories that the "nations were divided by their lands, each one with its own language, according to their clans, by their nations." (Genesis 10:5) Leibowitz sees the babel of languages not as a punishment but a corrective return to how things had been and were supposed to be.

That is still our challenge today. Diversity is not a liberal value; it is the way of the world. We know that the environment outside of our human lives is healthier with greater diversity, coral reefs and rainforests being prime examples. It is also true for humanity. We are better off because of the different religions, nations, cultures, and languages that comprise the human family.

But as that human family, we are shattering the web of diversity so essential to our environment and our lives. Moses, we are taught, had to respond to the shattered tablets that lay at his feet. At that point, it would have been easier for him to just walk away. But he did not. At the age of 80 he turned around, modelling *t'shuvah*, and climbed back up the mountain to get a new set of tablets. Like Moses we cannot afford to walk away from our task, no matter how difficult it may seem. We must return the diversity of the environment back to how it is supposed to be.

Rainbows

Margaret Frisch Klein

There are many songs about rainbows. One I learned at Girl Scout camp said that I could even sing a rainbow!. Since that summer, rainbows have had a special place in my heart.

Learning the story of Noah in a *parshat hashavua* class in college was one of those moments. In order for a rainbow to appear, there has to be a perfect balance between sun and rain. Without that balance, no rainbow. Without G-d and a certain balance, no world. For me, in that instant, G-d was perfection. So what would be more appropriate than G-d choosing the

---

[1] Leibowitz, Yeshayahu. *Notes and Remarks on the Weekly Parashah*; Israel: Chemed Books, 1990, pp. 17-20.

rainbow as the sign that G-d would never destroy the world again by flood? Rain is good. Too much rain leads to flooding and destruction.

Rainbows have appeared at my daughter's Bat Mitzvah, my daughter's high school graduation, my installation as rabbi. They are often a quiet, private reminder that the world is good and G-d is in charge. An almost private blessing, usually when I need that reassurance the most. What I have discovered is that you can't go chasing them. You need to be surprised by them.

One Rosh Hodesh Av, a day when we begin to feel the sadness of the impending Tisha B'Av, I was surprised by a rainbow. It was particularly surprising because the night before I had gone looking for one and didn't find it. Remember, it is about that perfect balance. This one was a beautiful, exquisite, full, double rainbow over Lake Charlevoix, in northern Michigan. You could actually see both ends reflecting in the water. Now the slogan for Charlevoix is, "Once in a blue moon there is a Charlevoix." This was not a blue moon. This was even rarer. Cars stopped on both sides of the road to take pictures. Everyone said, "Wow." It was a holy moment—regardless of religious tradition.

Sometimes you can capture such images on your camera. Other times you cannot. Sometimes you are surprised. On a hiking trip to Starved Rock State Park to prepare for Rosh Hashanah, the trail sign said, "Return." Perfect for Rosh Hashanah, I thought, and its themes of *t'shuvah*, return. I snapped a picture. When I returned, I found that the sunlight had caused a rainbow in the picture just over the trail sign.

Each time rainbows appear, they are a reminder of G-d's covenant with us. Each time, they are an opportunity to recite the blessing, "*zocher habrit*," that G-d also remembers the covenant with us, is faithful and keeps G-d's promise. Each time, I wonder how I keep my end of that covenant. How do I make good on the promise to be a partner in G-d's creation, to never destroy the world? How do I leave this world better for my children and my children's children? The rainbow, while offering that reassurance of G-d's love and promise, also demands an answer to that essential question.

Baruch ata Ado-nai Elo-heinu melech ha'olam zocher ha'brit v'ne'eman bivrito v'kayam b'ma'amaro. Blessed are You, Adonai our G-d, Ruler of the Universe, who remembers in faithfulness the covenant and the promise to keep the covenant forever.

Elul 17

Sunrise, Sunset—Evening the Frayed Edges of Our Lives

Jeff Foust

Sunrise and sunset are special liminal times calling forth awe and mindful awakening to spiritual realities we otherwise might totally miss. It's no accident that the main traditional prayer times for Jews are sunrise and sunset.

This simple profound reality is especially moving me this year as I prepare for the *T'shuvah* (Realignment/Renewal) work of Elul before Rosh HaShanah. I've been reflecting on a powerful liturgical adaptation by Rabbi Geela Rayzel Raphael[1] of the opening evening prayer Ma'ariv Aravim. She calls it "Evening the Evenings." It combines interpretive English with the traditional Hebrew. What especially moves me is the chorus: "Evening the evenings; evening the frayed edges of our lives; *Ma'ariv aravim…*; amen." The key for me is that "*Ma'ariv aravim*" refers both to "The One Who brings on the evening" and to the Creator of the heavenly vaults of light and darkness (*Aruv* can be a heavenly vault or a containing boundary) which almost come together at the time of sunrise and sunset (in Hebrew called "*Bein HaArbayyim*," Between the Heavenly Vaults of Light and Darkness).

When I hear and experience "Evening the evenings," I experience the light of "*Bein HaArbayyim*" touching the shadows and ragged edges first between the heavenly vaults of day and night at sunset, and then touching all the shadowy, dark, constricted, frayed places of the world, including in my own self and my entire bodyheartmindspirit. I actually tremble and shake in the original sense of *haredim* (the tremblers), but it feels good because I know that the path to *t'shuvah* and renewal is in letting my frayed, shadowy dark place be evened out by G!D's light and love.

With G!D's help and our sincere efforts, may our Elul journeys be enlightening and renewing.

---

[1] See www.shechinah.com for more of Rabbi Raphael's work.

Reconnecting with the Lands of My Birth

Lawrence Troster

On a vacation some time ago to my home town of Toronto, I drove around the countryside and saw the many places I knew so well from my childhood. As I drove, I reflected on how the landscape in which I lived affected who I am and how I see the world. The Toronto area was covered by glaciers over 10,000 years ago, and the land still is shaped by that ancient event: spoon-shaped hills called drumlins, ridges called eskers, which are the remains of the river beds that flowed from the retreating ice. And lakes: I spent many of my summers at camp in Northern Ontario beyond the glacial till where the major geological feature is the Canadian Shield. It has some of the oldest rock in the world—more than 3.96 billion years old—and covers some 5,000,000 square miles of Canada and the United States. The glaciers carved out more than 250,000 lakes from the Shield, and many of my summers were spent in the rock, water and forest of that landscape. In this world, I had some of my deepest and most important spiritual experiences in the Canadian Shield.

When I moved to New Jersey over 20 years ago, I became part of a new geological area: the Piedmont Province formed of volcanic basalt over 200 million years ago. And now I live in Pennsylvania, where I am in a different kind of geological formation: the rolling hills and valleys of metamorphic rock formed during the Precambrian period some one billion years ago.

What does all this mean for me? The food that I ate and which formed me was grown in the glacial soil of Southern Ontario and is still, so to speak, bred in my bones. How did this land also affect my mental perspective on the world? I thought of these things as I saw the familiar ridge of the Niagara Escarpment over where the mighty Niagara Falls is found. For the first time in my life, I took the boat that brought me close to the Falls. I felt the spray and saw the wonder of those thundering waters.

In this month of Elul when we are supposed to engage in *heshbon ha-nefesh*, to take stock of our lives and actions from the past year, I believe that we should also think about the places where we were formed and where we now live. Forgetting these landscapes is a kind of sin. We must remember the rocks, the soils, the water, the flora and the fauna and what they imparted and continue to impart to our lives in real concrete ways. Each one is different; each one has special qualities that we are mostly not conscious of. So as part of our spiritual accounting we should try to bring these places out of our unconsciousness into our waking thoughts. Maybe this process will teach us to understand how we are of the Earth and not just the consumers of earthly things. In a real way, as we live our lives, we are reborn from the places on Earth on which we dwell.

Our tradition often tries to symbolically connect us with the Land of Israel which provides a foundation for our identities as Jews. Collectively, it is the land that formed us as a people and where we still live in our collective memory. But each of us also has a place and a foundation where we were born and nourished by the gifts of the whole Earth. It is the true home from which we came, an actual place where the minerals of the soil, the water we drank, and the air we breathed have given shape to our flesh. Let us not forget these places. Let us remember and ask for forgiveness for the sin of forgetting the places from which we came, on which we dwell, and to which we will go.

Elul 18

Remembering Earth

Steph Zabel

"I thought the earth / remembered me, she / took me back so tenderly, arranging / her dark skirts, her pockets / full of lichens and seeds."[1]

This line from Mary Oliver's beloved poem, "Sleeping in the Forest," often runs through my mind. Especially when I leave behind my city environs and return to the embrace of the forest and green, wild places.

T'shuvah, return.

Some of us may be more drawn to the outdoors than others, but I believe that each of us has heard a call to return to nature at some point in our life. A return to nature can simply be a momentary remembrance, a moment of connection and acknowledgment. Even the most citified person, with feet always on concrete, can be stopped in their tracks by the pink glow of a sunset, the tremendous power of a thunderstorm, or the pure beauty of a newly opened flower.

Nature calls to us in so many ways—in many languages and colors and scents and whispers. It is simply up to us to be aware of these communications.

For me, I most strongly hear how the Earth speaks through the beauty of plants. I see the divinity of all life reflected in the body of plants. The flowers, leaves, seeds, and roots of these beings contain healing for our own bodies and spirits.

I see this in the way a flower blooms and then sets its seeds, in the changing colors of autumn leaves, in the waxing and waning rhythms of the seasons, in the abundance and diversity of life that greets us everywhere. The scent of a flower, the sunlight filtering through the leaves of the trees, the feel of the grass underneath my bare feet, the wind on my face, all return me to the Source of everything.

The Earth will always remember us and will endlessly call us back to her. We must simply listen. And then, we will remember the Earth.

---

[1] Oliver, Mary. (1993). "Sleeping in the forest." *New and Selected Poems*. Beacon Press, 1993.

Spiritual Charity and the Tale of Two Seas

Ziona Zelazo

During a summer stay in Israel, I heard from my friend Dalia about her nephew, who was killed in a terrorist attack. In his death, he donated his organs to save lives. He enabled a man to regain his vision with a donated retina. I was thinking how amazing it is to be able to give to others. But in particular, I was thinking that there is no one way to give to others. People can choose to be givers in many shapes and forms.

And here is another Israeli hint for the idea of giving: There are two lakes in Israel. One is the Dead Sea, the other is the Sea of Galilee. Both are not really seas; both receive their waters from the Jordan River. And yet, they are very, very different. The Dead Sea, in the south, is very high in salt. You can float and read a book at the same time! Thus, there is no life at all; no vegetation and no marine life. Hence the name: Dead Sea. The Sea of Galilee, in the north, is surrounded by rich and colorful vegetation. It is the home to over twenty different types of fishes.

Same source of the seas' water, and yet one is full of life, the other is dead. Why? The Jordan River flows into the Sea of Galilee and then flows out, so it keeps the sea healthy and vibrant, allowing aquatic life to exist. But the Dead Sea is below the mean sea level and has no outlet for its water. The water flows in from the Jordan River, but does not flow out. Thus, unfit for any aquatic life.

There are the Givers and the Takers. And in Judaism, giving charity is an obligation. In the Bavli Talmud, we find dozens of texts about this obligation. They say that *tzedakah* (charity) is the most important commandment to fulfill. For example, we read in *BT Bava Batra 10a*:

> R. Yehudah says: Ten strong things have been created in the world. The rock is hard, but the iron cleaves it. The iron is hard, but the fire softens it. The fire is hard, but the water quenches it. The water is strong, but the clouds bear it. The clouds are strong, but the wind scatters them. The wind is strong, but the body bears it. The body is strong, but fear crushes it. Fear is strong, but wine banishes it. Wine is strong, but sleep works it off. Death is stronger than all, but charity saves from death, as it is written, Righteousness *tzedakah* delivers from death (Proverbs 10:2; 11:4).

If you are like me, we were taught that *tzedakah* is giving money to the poor or to other worthy causes. But that misses the point. Most often, we think that giving relates to how much money one gives away. That if we donate to charity with a check, we are Givers. And if we do not have the means of money, we are out of the box of Giving. However, being generous is far more than a money issue. It is a code of behavior that requires generosity from the heart, the sharing of personal time, energy, talent, wisdom, love, compassion and many other resources. The act of charity also includes visiting the sick, burying the dead, and dealing justly with others.

The classic ethical work *Orchot Tzadikim: The Ways of the Righteous*, written in the 15th century, liberates us from this conditional relationship between charity and money. Three categories of giving with generosity are listed: giving of one's wealth, giving of oneself physically by being present to others in need, and giving of one's wisdom. The last two are types of giving that money cannot buy.

The giving from our spirit of compassion and love is the emotional quality of giving which is commonly referred to as *gemilut hasadim* (deeds of loving-kindness). Investing from your own energy builds relationship, which does not depend on giving money to the poor. In *BT Sukkah 49b*, we read:

> R. Elazar further stated: Acts of loving-kindness (*gemilut hasadim*) are greater than charity (*tzedakah*), for it is said, 'Sow to yourselves according to your charity (*tzedakah*), but reap according to your *hesed* (kindness)' (Hosea 10:12); when one sows, it is doubtful whether he will eat [the harvest] or not, but when one reaps, he will certainly eat.

Our rabbis taught: In three respects, *gemilut hasadim* is superior to charity: charity can be done only with one's money, but *gemilut hasadim* can be done with one's person and one's money. Charity can be given only to the poor, but *gemilut hasadim* both to the rich and the poor. Charity can be given to the living only, *gemilut hasadim* can be done both for the living and for the dead.

Have you considered visiting the sick as giving? Or, supporting someone by just listening to them without trying to fix them? Could we imagine that a smile, a recognition of someone with words could also be acts of giving? *Bava Batra 9b* affirms that our presence can lift another's spirits at times of despair and sustains the recipient at least as much as any donation:

> Someone who gives a coin to the poor will be blessed with six blessings, whereas the one who addresses him with words of comfort will be blessed with eleven blessings (even if he does not give him a donation).

I think that each one of us was, one time or another, the Giver of many versions, like the Sea of Galilee, where it enriched our lives with the sense of a moral fulfillment. And also being the Taker, where it left us spiritually empty. How do you see yourself today? Are you the Sea of Galilee or the Dead Sea?

Let's not be like the Dead Sea. May we recall the joy we get when giving and what healing it can provide because the act of giving from the heart makes a difference in people's lives.

Elul 19

Time to Rest

Tali Anisfeld

Night summoned day again this morning, and the dawn told me to rest. This surprised me. Isn't it the setting sun—rather than its rising companion—that coaxes us into rest? But then again, maybe there is something about rest that comes with motion, rather than with stillness. The reach and stretch of the day rather than the hushed retreat of night.

It is time to rest—time to awaken in prayer and laugh with the tickling grasses. To climb and jump and run and let the sweat of your body mix with the rain of the heavens. It is time to listen to music without singing along. To think not about the words or the notes, but the way it feels when an entire universe splashes over you and dives down, down, down. Chest. Belly. The soles of your feet. It is time to rest.

Look at that old tree. A tree that has stood, still but trembling. Tell it your secrets. I'm tired. I want to dance and sing, but sometimes my legs weaken and my voice gets hoarse. How do you do it, old tree? Are you tired? Do you need to rest your weary legs and heavy head, too? Not even in the soft ache of a cold winter night?

I recently fell in love with the liminality of tidelands. Nowhere in the world does change make more sense. Ebb and flow—come and go, I need you, you need me—the ocean's mighty beckoning and gentle retreat. And a whole world of creatures. Clutching desperately yet calmly—not in fear, but in something related—to the Earth moving beneath them. Needing water, but not only water—and land, but not only land.

Maybe when the trees quiver in the cold and rustle in the wind, they're resting. Maybe staying in one place—standing straight and steady and still—is harder than bending over or craning backwards, stretching until you feel a muscle you'd forgotten. Because to stretch is to breathe is to move is to grow stronger.

Chant along with the beating rain. Gaze on the mystery of radishes growing in the yard. Shiver with the frog on the bank of the river that has lost his family. Weep for the flowers that wither too soon and rejoice with the morning songbirds. Let them do the same with you. Even let them do it for you sometimes.

T'shuvah is rest in motion.

To grow as a tree grows—to return to the Earth as you stretch up, up, up. *Shuv*.

And don't forget there are blueberries growing up in the mountains, and the world is created each morning, and so are we.

Walking Prayer; Dancing *T'shuvah*

David Arfa

> "For many of us, the march from Selma to Montgomery was about protest and prayer….Even without words, our march was worship. I felt my legs were praying." —Rabbi Abraham Joshua Heschel[1]

Yes, our legs pray. Not just when marching either. Our legs, our entire body and all of our words are continuously interwoven in deep connection with the entire cosmos. Rabbi Heschel the Kabbalist may have been shy sharing this full power of Judaism's mythological grandeur. For me, Judaism's mythological grandeur, beauty and deep integration with Jewish practices have kept me Jewish. How can this ancient mythological power be activated in a single protest, a single dance? Listen in.

In the ancient days,[2] when midrash was the cutting-edge commentary of the day, our Rabbis taught about the secret holiness of our bodies. They said that every commandment has a corresponding place in our body and a corresponding day of the year. Because of this, we can carry the holy commandments (which come from the Holy Torah and Gd the Source of All) in our bodies and days. This is a powerfully bold take on being created in the image of Gd.

This teaching was carried forward into Medieval Kabbalah where the Kabbalists actually taught that the entire cosmos needed our actions, our prayers for Gd's own healing. Our actions close the loop, and we participate in an abundant and interconnected flow of blessing continuously falling from the 'heavenly realms' and raised back up again, being replenished by our actions and prayers. A sacred spiritual ecology, if you will. We are not only saying, 'As above, so below', but also, 'As below, so above.'

Our wonderfully creative teachers from the 1700's transformed this teaching by adding everyday dance (song, story and plain words too!) to the list of powerful spiritual recycling tools that help 'unify the unities.' Did you

---

[1] Retrieved from http://www.dartmouth.edu/~vox/0405/0404/heschel.html

[2] For additional background on the powers of dance see Fishbane, Michael. "The Mystery of Dance According to Reb Nachman of Bratzlav." *The Exegetical Imagination: On Jewish Thought and Theology,* Harvard University Press, 1998.

know the Shpoler Zeide continued to dance with the lightness of youth well into his old age?[1] Once, a Jewish life was in danger. A giant Cossack soldier was cruelly treating him like a cat treats a mouse. The giant declared that if anyone could out-dance him, then he would spare the life of this simple Jew. However, if not, then both dancer and hostage would die! Everyone was so scared. It was the grandfatherly Shpoler Zeide who stepped forward. He danced the Bear Dance with such focused power and vigor that the Cossack was unable to keep up. He fell down laughing saying, 'You win old man, you win.' For the Shpoler Zeide, dance was a superpower! Able to affect *t'shuvah* with a single bound.

Other times, when disputes would reign, the Shpoler Zeide would dance the Heron Dance, slow and graceful, and conflict would melt away. So much so, people would say that the Shpoler Zeide was a master of dance and able to achieve Holy Unifications with each step of his foot!

Reb Nachman of Bratzlav, another holy teacher from that generation, actually prescribed dance as a remedy for the hopeless despair that prevents joy. He knew the act of dance was enough to raise joy high in the saddest of souls. Dance as medicine. How's that for creative health care!

This is Rabbi Heschel's legacy. He's not just protesting, he's praying with his legs—adding healing unifications for the world and for all of us. And his powerful feet are still inspiring us today! Inspiring us today to take a stand in the streets with our neighbors. Reminding us of the power we hold in our legs, in our actions. Reminding us to appreciate the miracle that we have legs to stand, ankles that rotate, 26 humble bones of the foot that allow us to stand steady even on uneven ground.

This Elul, let's bring all of the enchantment we can muster to our prayer, to our *t'shuvah*. Let's remember the spontaneous freedom of prayerful dance that we carry and its ability to mark a new path. We truly can alter the shape of tomorrow. As Emma Goldman is known as saying, "If I can't dance, I don't want your revolution!"[2] Rally Ho!

---

[1] Stories adapted from Buber, Martin. *Tales of the Hasidim*. Schocken, 1991.
[2] http://www.lib.berkeley.edu/goldman/Features/danceswithfeminists.html

Elul 20

Returning to Our Pond

Dorit Edut

Frozen for months, life had chilled out for too long last winter. We began to wonder if a new Ice Age was coming more swiftly than predicted. Disaster was whispering in the wind from which we tried to hide all skin lest the frost take a bite. We stayed indoors and cancelled many a get-together because of the fierceness of this weather. On the pond in front of my daughter's home, the white heron appeared once in March as if sent by Noah, but all was solid ice.

On early April evenings, the story of our Exodus from slavery to freedom was told and we began to feel again in our fingers, toes, and inner recesses the need to move forward, to act with the courage of our ancestors and confront our world. This time the heron found a patch of water amidst the melting ice and it, too, looked relieved.

Spring rains brought flooding waters in some places but a rainbow of colorful flowers appeared then in June and July gardens—and we knew that our reprieve had been granted once again. Geese and ducks glided by in the shaded waters of the pond, with newborns in tow. We invited friends over to join in feasts of ripe cucumbers, tomatoes, eggplants, beans, corn, peaches, berries, all sprinkled with fragrant herbs from our abundant gardens. The summer pleasures encourage healing and strengthening of bodies, minds, and spirits, until they have almost reached a stage of perfection.

Yet, then it is that we also start to turn towards fall, the time of our New Year, the ingathering of harvested crops and friends and new projects—for winter is sure to come again and no one knows how long it will last this time or who we will have become when the heron returns next spring. We look now and see our reflections in the pond's waters and quietly pray with a full heart, saying thanks for all this and asking or hoping for more.

In the floating leaves which appear on the water's surface, we are aware that another year has passed, and we have left much to be done to repair our world. Can we stave off winter? Perhaps not, but we may prevail over its excesses by the warmth of deeds of lovingkindness, reaching out to those who stand on the shores of other ponds, big and small, where our white heron has also flown.

The Humility of Rabbits

Leora Mallach

I am an educational vegetable gardener, in that I facilitate learning about food and grow vegetables for people to eat. Vegetable gardens don't happen by chance but are manicured and maintained on a regular basis. There is pre-season planning, worry and hope as things sprout, groups of students to program with, volunteers to direct and family picnics to coordinate.

When I first saw evidence of the rabbits over the winter, I didn't totally understand the implications. Ever the optimist, I thought they could hang out in the ivy, frolic in the playground (once the pre-school kids left) and generally leave me and my vegetables alone. They could have their space, and I'd have mine.

And then they ate my pea shoots.

Planting continued on, and weeks later, who was I to tell the 9th grade boys who relished in cutting back ivy that the cute little rabbits had eaten up all their hard work? Or the volunteer who had gotten so excited when she planted her first seeds (Cherokee Wax and Purple Queen beans) that she took a picture of the patch of soil. Would she want to know that her beans were now mere colorful sticks?

I raised my fists at the rabbits. I cursed at them. I chased them. They shook their white bunny tails at me and scurried away.

I called the experts, some helpful and some not. Fencing would have to be dug six inches down. How would the beds still be accessible? Would they ruin the aesthetic of the space? Wait it out some said, once the plants are big enough, rabbits won't want to eat them. An exterminator would use a gas chamber… uh, NO.

The advice I went with was to become a rabbit harasser. I sprinkled fox urine around the rabbit hole so they would think they were being stalked. I had friends bring over their dogs to "leave their scent" in the area. I sprinkled bovine blood granules on the beds next to the vegetables.

When I could take it no more—I bought a trap. The first morning when I went to go check the trap, I wasn't sure if I really wanted to find a rabbit inside. I had located a lovely new home (more than two miles away, by the water, which included a bridge and a bench in addition to a wide grassy area). But I was nervous. If I hold their lives as sacred and their creation as an act of divinity, then shouldn't we be able to coexist in the garden together?

I often wonder what these rabbis are teaching me. I am still learning. In this time of *t'shuvah*, of love, of renewal, of working toward our best selves, I am humbled by the rabbits.

Elul: A Time to Start Shifting Our Imperiled Planet onto a Sustainable Path

Richard H. Schwartz

Elul is here. It represents an opportunity for heightened introspection, a chance to consider *t'shuvah*, changes in our lives, before the Days of Awe, the days of judgment, the High Holidays of Rosh Hashanah and Yom Kippur. The shofar is blown every morning (except on Shabbat) in synagogues during the month of Elul to awaken us from slumber, to remind us to consider where we are in our lives and to urge us to consider positive changes.

How should we respond to Elul today? How should we respond when we hear reports almost daily of severe, often record-breaking, heat waves, droughts, wildfires, floods, and storms; when every decade since the 1970s has been warmer than the previous decade; when the past four years are the warmest four years since temperature records were kept in 1880; when polar ice caps and glaciers are melting faster than the worst case projections of climate experts; when climate experts are warning that we could be close to a tipping point when climate change will spiral out of control with disastrous consequences, unless major changes are soon made; when we appear to also be on the brink of major food, water, and energy scarcities; and when, despite all of the above, so many people are in denial, and in effect, seem to be "rearranging the deck chairs on the Titanic as we approach a giant iceberg"?

It's well known that one is not to shout fire in a crowded theater. Except if there actually is a fire. And, the examples indicate that the world is on fire. Therefore, we should make it a priority to do all that we can to awaken the world to the dangers and the urgency of doing everything possible to shift onto a sustainable path. We should urge that *tikkun olam* (the healing and repair of the world) be a central focus in all aspects of Jewish life today.

We should contact rabbis, Jewish educators, and other Jewish leaders and ask that they increase awareness of the threats and how Jewish teachings can be applied to avert impending disasters. We should write letters to editors, call talk shows, question politicians, and in every other way possible, stress that we can't continue the policies that have been so disastrous. We should make changes in our own lives to reduce our carbon footprint by driving less, recycling more, eating less meat, and more.

The afternoon service for Yom Kippur includes the book of Jonah, who was sent by God to Nineveh to urge the people to repent and change their evil ways in order to avoid their destruction. Today the whole world is Nineveh, in danger of annihilation and in need of repentance and redemption, and each one of us must be a Jonah, with a mission to warn the world that it must turn from greed, injustice, and idolatry, so that we can avoid a global catastrophe.

Elul 21

Community and Covenant

Katy Z. Allen

I've been thinking a lot lately about community and covenant.

Rabbi Avi Olitzky defines community as "a circle to which you feel you belong that will miss your presence; it reaches out to you when you're absent, and you long for it when you're not there."[1]

A *brit* (covenant) is a promise, generally bilateral, that requires the participation of both parties bound by the covenant.

In the Torah, G!d enacts three covenants. The first is G!d's promise to all humanity after the Flood never again to wreak such destruction. The sign of this covenant—actually a one-way agreement, because G!d promises unconditionally, and humanity is not obligated in return—is the rainbow.

The second is G!d's covenant with Abraham, promising to make numerous his descendants and to give them the Land of Israel for their possession (Genesis 17). *Brit milah* (circumcision) is the sign of Abraham's acceptance of the covenant and his loyalty to G!d.

The enactment of the third covenant takes place at Mount Sinai when G!d gives the Torah to the Israelites and outlines the terms of the covenant; this makes many demands upon the people. Shabbat is the sign of this last covenant.

These three covenants provide intimations about the three kinds of community to which we all belong.

The rainbow is part of the more-than-human world, outside of us, a reminder that community extends beyond humanity to the Universe and all it contains, both living and nonliving.

Circumcision is a reminder that our relationship with God is personal, that we must look inward to our hearts and our souls in order to fully maintain and grow our relationship with the Sacred.

Shabbat is a reminder to connect with humanity, to interact, celebrate, rejoice, remember, and observe together.

---

[1] Olitzky, Kerry M. and Avi S. Olitzky. *New Membership & Financial Alternatives for the American Synagogue: From Traditional Dues to Fair Share to Gifts from the Heart.* Jewish Lights Publishing, 2015.

Community is a circle to which you feel you belong that will miss your presence; it reaches out to you when you're absent, and you long for it when you're not there.

The more-than-human world—it calls out to us, but often we humans fail to hear its call. Yet, we are connected in our very DNA to all of life, and through stardust, we are connected beyond the living world to the nonliving world. When our hearts are open to its wonder, we long for the more-than-human world when we stay away too long.

Our hearts and our souls—they also call out to us, and when we ignore their commands, constantly, without rest, putting our nose to the grindstone of what must be done, our spirits shrivel and die.

Our human community—both our loved ones and those we have never met—call out to us in so many ways. We need each other, physically, spiritually, and emotionally—our very survival depends upon our being connected to each other.

To join together in community, to become one—one with the Universe, one with ourselves, and one with all humanity, to the best of our ability, this is our holy task on this Earth.

The journey through Elul and toward Rosh HaShanah invites us *l'chadesh* (to renew) our covenant with the Universe, with our souls, and with humanity, as we engage in *t'shuvah*, seeking to return to the heart of the three covenants and the three communities upon which we depend.

Natural Awe and Artistic Representations

Steve Altarescu

When the Israelites stood around the foot of Mount Sinai, the mountain appeared as if it were ablaze with fire, and they heard the sound of God from out of the fire, but they did not see any form or shape.

The experience of God without a form or shape taught us that making a likeness, a resemblance of anything in nature, would be disconnecting us from a direct experience of all that is sacred. One has to wonder though, why does Moses repeat this prohibition four times?!

One answer is that the Torah is teaching us that there is power to an image, whether it be a sculpture, a painting or any other art form that stands in contrast to feeling the power of God or God's creation.

I have found that there is a difference in the experience of being in the natural world versus seeing representations of what is in nature. Watching a hummingbird on our deck is very different from seeing the hummingbird engraved on my coffee mug!!! The hummingbird fluttering around has the power to open us up to an experience of great wonder and awe at the vast beauty, intricacy and inter-connectiveness of the natural world. Abraham Joshua Heschel wrote:

"The beginning of happiness lies in the understanding that life without wonder is not worth living. What we lack is not a will to believe but a will to wonder...The way to faith leads through act of wonder and radical amazement."[1]

There is an awesome beauty to creation, alive and always changing—when we walk outside and quiet the chatter of our minds and listen to the birds, the cicadas, or see a sunset, we can be blessed with moments of radical amazement.

The Torah teaches us to see this world and all of Creation as holy, sacred, and mysterious. We are challenged to look at the world and see the Divine Presence, not in any particular form or shape, but within the life energy of all forms and shapes. The rabbis of the Talmud acknowledge that humans will make and enjoy art. Even in ancient Israel synagogues were built with intricate designs and images of animals. I remember visiting synagogues built in Northern Israel 2000 years ago and seeing beautiful mosaic tile floors with all the signs of the zodiac.

The rabbis of the Talmud wrestle with the permissibility of artistic representations. One of the ways the Talmud resolves this question is to only prohibit making an image of anything that represents God or any of the items that were parts of the rituals of the Tabernacle in the desert or the Temple in Jerusalem. The reason given is that the Tabernacle and the Temple were the places for sanctioned artistic creation that God commanded.

Maimonides said that the purpose of the prohibition against copying the ritual art that God sanctioned is to promote reverence for the sacredness of that which would be lost if reproductions, copies of copies of items, are created.

Maimonides took the concept of reverence and awe for the objects of the Temple and related them to how one behaves in the Temple, including a particular sacred choreography performed by all the masses of people who attended a service. This choreography included many rows of people in circles who walked in circles that intersected with each other. One of the circle groups included those who were fortunate in life. They were instructed to follow a script in which they asked those in the other circle, the less fortunate ones, about their troubles and offered them words of comfort and strength. Like the cherubim in the Tabernacle, the space where they faced each other was where the *Shekinah* (the Divine Presence) would be revealed.

---

[1] Heschel, Abraham J. *God in Search of Man: A Philosophy of Judaism*. Farrar, Straus and Giroux, 1976, pp. 46-47.

I believe Maimonides wants us to understand that the experience of awe and wonder of the natural world will lead us to have reverence for what is most sacred: each other. When we face another human being with care and compassion, we have the opportunity to invoke the Divine Presence that is within and between us.

I suggest we begin the process of *t'shuvah* by looking inside ourselves while spending time in the quiet and open spaces of the natural world. If we start the process of *t'shuvah* with awe and wonder, we might see and grow the divine spark within and see this light in all we encounter.

Saying Farewell with Each Breath, Starting Anew

Judith Kummer

Towards end of the day, towards summer's end,
body and soul prepare for farewells.

Through piney woods I run,
gauntleted by trees whose dark limbs
reach up to breathe in fresh blue sky.
Dim path; the light can't reach down here.
Ahead, the river winks at me.

I thread my way out the wooded tunnel's end
and can feel the sky lift—and
my mind lifts too.
Before me lies
still water
meandering between wooded banks.

Turning, I race the river. Feet pound
on hard sand paths,
Pulse quickens in my ears,
breath pulses through my veins.
Crickets thrum in the long grass.
Feeling especially fleet of foot,
I dart between notes of birdsong.
My worries lag behind, can't run as fast.

And then,
pleasure-filled, I pause at river's bend
to glimpse reflected glory.

At golden hour
there's
stillness in the air
almost a hush,
a waiting for day to end night to come.

Time feels burnished,
and the light—oh, the light is still golden,
glowing.

Gild the clouds, light,
gild them colors of half the rainbow.
Mirror the light, river,
mirror the
awe-tinged clouds.

What a vision of beauty—
glory reflected in thinly silvered water.
I am filled up, full.

Brimming over
I turn back towards the start,
saying farewell with each breath,
ready to start anew.

Elul 22

Another Blue Day

Thea Iberall

I have a picture of my mother with Heidi Klum, the blonde supermodel and TV star. We were in Heidi's trailer on the Warner Brothers lot in Burbank watching her prep for a commercial shoot. Heidi and her makeup entourage gathered around my mother, who was wearing her "Kiss Me I'm 100" T-shirt. They wanted to know her secret to aging well. My mom told them about the gin-soaked raisins she eats every morning to ward off arthritis. Then she talked about the raw apple cider vinegar she takes before every meal to overcome gas. And the walnuts and blueberries and probiotics, the classes and crossword puzzles, how she plays bridge and Scrabble. And how she set a world record in swimming when she turned 90 years old.

My mom has lived a life of service, from the Campfire Girls to the National Council of Jewish Women. She tutored Russian immigrants in English as a second language. In 1974, at great risk, she smuggled letters and money to Russian Refusenik Jews in Moscow. In the middle of the night, she managed to avoid the KGB and meet with the scientist Alexander Lerner and also a young Natan Sharansky. At one of the apartments she visited, she was asked if she could speak Yiddish to an elderly Russian woman who had not heard the language in years. My mother agreed and they woke the woman up. Together, they cried and shared stories in Yiddish.

My mom died at the age of 102 at personal peace but upset with the world. To her, the only life worth living is one steeped in community and family. "Prejudice is the worst thing," she told me. Her mantra was, "Dwell on human kindness." As an Orthodox Jew, this was her Judaism. She said it to everyone she met, the young and the old. She also told them about her secret of aging well. About being with the Earth, not against it. She lived her *t'shuvah* by asking the same question every day, a question from the Thomas Carlyle poem "Today"[1] that she had memorized in the fourth grade: "So here hath been dawning / Another blue Day: / Think wilt thou let it / Slip uselessly away?"

---

[1] Carlyle, Thomas. *Critical and Miscellaneous Essays*, I. London: Chapman and Hall, 1872, pp. 292-93. Retrieved from tspace.library.utoronto.ca/html/1807/4350 /poem416.html

Re-Remembering Who We Are

David Jaffe

Born at home on a Shabbat morning, my son spent his first few hours on this planet snuggling against his mother's warm chest. One of the most striking visual images of that first day was the moment our midwife cut the umbilical cord that physically connected mother and child. Until that moment I knew abstractly that we were all connected and even, at rare times of spiritual reverie, sensed this connection. But here I saw it—as humans we were at one point actually physically connected to another human being, our life interdependent with their life! The loss of this raw, visceral sense of interconnection with all Creation may be the key psychic contributor to our human penchant for environmental destruction.

Mussar master Rabbi Shlomo Wolbe writes in his *Essays on Elul* (Elul 1958):

> "The entire creation is unified and clings together, for 'We all have one ancestor,' and all creation draws close to one another, one great family. We humans are close to the inanimate world, for it is written, 'We are dust.' We are close to plant life for we also have the life force in us. Our closeness to animals is even more pronounced. We don't even need to say how much closeness there is between different nations and races...."[1]

For Rav Wolbe, connection comes from shared properties. We are made of minerals and water so we are intrinsically connected to the physical Universe; we grow so we are connected to all that grows. Why then would we act in such destructive ways towards the planet, animals and other humans? Rav Wolbe points out that the root of the Hebrew word for cruel *akhzar* means estrangement. Only when we make other people or the Earth as strangers can we be cruel. This estrangement comes from our confusion about who we really are and what we are doing on this planet. Elul is the season for correcting this mistaken sense of distance and alienation. It is the time to literally re-member who we are and how we started. Just like my son with his umbilical cord intact, we are deeply connected in a real way with the people around us, with the Earth under us, the sky above us, and the divine soul within us. May knowing this end our destruction of this planet and all its inhabitants.

---

[1] Wolbe, Shlomo. The Forces of Amity and Alienation in the World. In *Essays in the Days of Desire.*

Enfold me, Earth

Carol C. Reiman

Enfold me, Earth,
entwine your thick limbs
with mine;
lift me up above
your blushing beauty,
opening me to your new day.

Show me how to know you
as we whisper in each other's ear—
willow rustle,
sizzling spray upon the sand;
how I meant to help,
how I hurt you,
how we can heal
in easy forgiveness,
how I can keep you as
you keep me—whole.

Dance in freedom,
moving together,
none to crowd out
both our voices—yours and mine—
gulls' cry, sudden thunder;
rushing torrents, oaks riven
into fresh surfaces
for new growth.

Groom me to your model,
shaker of change,
mentor of shores that welcome
both fresh and salt;
teach me to leave
my shell behind,
for other use
as home for hermit crab.

Reach for fresh tastes,
those not yet sampled;
urge your passion onward,

twist, turn, recombine.
yet hold me, Earth,
in your warm embrace,
as I seek to reach
the Consciousness
of Creation
in the New Year.

Elul 23

The Compost Bins in Our Heart

Katy Z. Allen

My compost bins have always been much more than just a place where compost happens. The area beside the three wire and wood bins is a place where I have often felt my father's spirit. He was raised on a farm, and though he became a professional, gardening was in his blood, and he spent much of his spare time in his garden and his orchard.

Yet, it is not just the reminders of my father or the sense of his hovering spirit that have given meaning to my compost bins over the years. They are symbolic of so much—which may be the truer reason that I often think of my father when I take out the compost.

We gardeners deposit plant food wastes, garden trimmings, paper towels, chopped up leaves, and other materials into our compost bins. We let the rains add water, and from time to time we add a bit of soil and turn over the decaying material. Then we let nature take its course, and before too long, all of that "waste" has turned into dark, crumbly humus that will enrich the soil of our garden. The leaves, the banana and orange peels, the corn husks—all this and so much more have been transformed from something seemingly useless, a by-product, into something good, useful, and enriching.

When my heart is feeling heavy, and I sit quietly beside my compost bins, I, too, get transformed. The grief and sadness in my heart are lifted, and I find myself once again able to be useful, to myself and to others. I am able to forge ahead into new territory. My relationship with the Holy One of Blessing has deepened.

This, in essence, is what *t'shuvah* is about, turning the excess materials of our hearts and souls—those feelings of sadness, anger, jealousy, and more—into a deeper and closer relationship with G!d—re-turning to G!d—and in the process finding ourselves enriched.

It has been, I believe, in large part through my connection with my father, who passed away more than 40 years ago, that I have learned to grieve and to be transformed. But grief is complex, it is not a one-time endeavor; it is a mosaic of many feelings and experiences, and it returns, often when we least expect it. It shows up in new ways in response to new losses, so that frequently throughout our lives, something new and different needs to be transformed.

Thus it is for all of us, and thus it is in life. And so, our tradition provides the vehicle of the month of Elul leading up to Rosh HaShanah and all the days of the High Holy Days, to give us the opportunity to let our compost be transformed, to let our grief, our fear, and our despair be released, and to let our hearts open wider, in an ever-deepening relationship with the Mystery That Is.

Compost happens. May our transformation also happen.

Searching for the Tree of Heaven

Rachel Aronson

Despite its nickname, "the tree of heaven," the *Ailanthus* is not universally beloved. It is not planted in garden beds, on streets, or in parks. There are 22 types of permitted street trees in New York City, where I live, and the Tree of Heaven is not one of them.

The Tree of Heaven is most famous for being the titular tree that grows in Brooklyn:

"Some people called it the Tree of Heaven. No matter where its seed falls, it made a tree which struggled to reach the sky. It grew in boarded-up lots and out of neglected rubbish heaps... It would be considered beautiful except that there are too many of it."[1]

As a fan of *A Tree Grows in Brooklyn* and an amateur urban naturalist, I resolved to find the tree when I moved to New York. My search was, at first, entirely unsuccessful. The majestic trees lining brownstone streets were honey locust, oak, but never *Ailanthus*. Bike rides down tree-lined corridors found planes and tulip trees, but never *Ailanthus*.

A bit of googling shed some light on my problem: the *Ailanthus* is considered a weed tree. Quickly growing, with pods that produce millions of seeds, it's the tree equivalent of a dandelion.

So, I started looking for the *Ailanthus* in places where trees aren't planted. And I found them: in the middle of the subway tracks, growing out of abandoned lots, on uncultivated roadsides. Where no money had entered to beautify or to plant, there was the *Ailanthus*.

A friend of mine recently relayed a Midrash about Moshe and the burning bush. To find the leader of the Jewish people, God set up a fire in a bush that was not consumed. Shepherds came and went, their minds on other things, and overlooked it. Moshe was the first to see the bush for what it

---

[1] Smith, Betty. *A Tree Grows in Brooklyn*. 1943. Harper Perennial Modern Classics, 2005.

was—a miracle. And for this observation and appreciation, he was chosen as a great leader.

If left unchecked, the *Ailanthus* has been known to wreak havoc. It's an invasive species; its roots overtake sewer systems, its branches intercept telephone lines. I am not advocating for an end to thoughtful land management practices. Simply an appreciation of what is around us, a reminder to notice. To notice not just the beauty that's obvious before us, but to pause and see the beauty that we've been overlooking—that which might be considered a weed.

What will you notice today?

Elul 24

Seeing the Beauty

Sandra N. Daitch

I am grateful for the opportunity I had to spend a week out West with my brother and his family where I was gifted with seeing and experiencing the magnificence and beauty in nature. I saw Muir Woods, Scenic Coastal Route 1 in California, the Grand Canyon, and the red hills in Sedona, Arizona.

In those places, it was easy to feel the awe and joy of the Universe. Back home, in my Boston-area suburban apartment with all my things taking up space, I feel more challenged to stay in touch with the beauty and spaciousness of the Earth.

Looking outside my bedroom window, on a very hot humid day, I see a familiar, beautiful, large tree with leaves and branches moving in a gentle wind. I've lived here for over 15 years and always thought the tree was a maple. But in looking at the leaves now, I realize it's not. I'm not sure what it is, so I'm excited to have a chance to learn something new about this familiar lovely tree friend.

I've been away and my garden has lots of weeds. I can turn away with displeasure, or I can choose to appreciate the fertility of the soil that allows those weeds to grow. Some even have flowers. What is it about weeds that give them a bad name? They are wild, uncultivated beauty—the gifts that come without asking. Perhaps, it's just a matter of perspective?

Looking out through the window of my study, I see a number of trees and bushes in the neighborhood. I'm enjoying the variety of hues of green, and the distant tree with purple leaves—maybe a red maple? I'm also enjoying the shadows that shift under the tree closest to my view. There are also houses and several telephone wires outside the window, and I ask myself how to see the beauty in them. A thought occurs that I can use the wires to look at slices of the scenes—looking at the scene between each two wires. There are so many options, if I pause and ask what's possible in each situation. And I actually like the colors of the white and red houses within view.

There is so much beauty and good in the world, and, I get caught up in my daily chores and issues and forget to notice and appreciate the good and beautiful that's in my life. Going on vacation to beautiful spots in nature is a reminder to *t'shuvah* (return) to a practice of regularly taking a stance of gratitude, awe, and appreciation for our many gifts.

Collective Versus Personal Action in the Jewish Bible

Andy Oram

Environmental activists are constantly juggling between the personal and the political. Do we devote our efforts to using our cars less, substituting vegan meals for meat, and recycling? Or do we canvas our friends and neighbors to pressure governments and businesses to adopt more planet-friendly technologies? We know that we need to do both the personal and the political, but those who have taken the environment as our cause have found ourselves swinging between them in a way that is frustrating and distracting. And as we prepare for the High Holidays, we always look for how to do more good in the upcoming year.

Perhaps we can learn something from the historical experience of the Jews. As a *kehilah* (community), we have constantly explored the relationship between personal responsibility and communal action. Many High Holiday prayers, such as *Al Chet* and *Ashamnu*, refer to the community in the plural even though the sins must be addressed by each individual on her own. The twice-daily *V'ahavta* prayer shifts abruptly (Deuteronomy 11:13-21) from the singular "you" when prescribing behavior to the plural "you" when describing the positive or negative outcomes of this behavior: rain and food at the proper times versus drought that drives us from the land.

The grammatical shift suggests that each of us must take personal action to preserve the Earth, while the results will affect all of us irrespective of our roles in creating environmental damage. And the truth of this observation is visible throughout the world, as people with small carbon footprints get deprived of their livelihoods by climate change and leave their homes to suffer war or to deteriorate in refugee camps.

So, Jews understand that personal concerns are also communal ones. But the record becomes muddier when we look at the history of "people power" in Israel. In fact, the Bible gives us little to celebrate. Communal Israelite acts include the idolatry of the golden calf (Exodus 32:1-6), the invitation to the Benjaminite men to replenish their tribe by abducting women from a religious festival (Judges 21:20-23), and the demand for a king (I Samuel 8:4-22). The leaders of the Israelites concur in all these disastrous decisions.

To find a positive example of the relationship between policy and individual action, turn to the evil city of Nineveh in the book of Jonah. After the reluctant prophet proclaims the destruction of the city, the people of Nineveh, "from great to small," take penance on themselves (Jonah 3:5). Upon hearing of the prophecy, the king joins them and declares the spontaneous fast to be a policy. Sackcloth and ashes here represent both a personal sacrifice and a public statement, like building a solar farm and then pressuring the government to connect other people to it for electricity.

When we want to change behavior, we should start with ourselves. But we need not be so ascetic as to hamper our beneficial efforts. For instance, environmental leader Bill McKibben has assured followers that taking an airplane to attend a climate change rally is a good expenditure of carbon—the best, in fact.

If we persuade friends and religious congregants to change their individual behavior, we can also transform them politically. After putting hours of effort into composting or taking public transportation, a person naturally starts to think, "What if another hundred million people could do what I have done?" This should lead them to investigate the structural barriers that keep others trapped in environmentally damaging lives and to demand political changes that spread the good they've done even further.

Like all deep and abiding social changes, the shift to sustainable human life will be a grassroots movement that blossoms into political action.

Circling Home

Kaya Stern-Kaufman

Turning, always turning
To every turn, a time
To every time, a spirit
To every spirit, a soul
To every soul, a home

Turn as the earth
Around your sacred truth,
Hold to your center
     but move from your place.
See
  from a new view, who You Are
And who you need to be

We are fiery emotions
We are waters of compassion
We are centered breath of life,
We are steady solid clay

An eternal breath wrapped in dust and light
Turning
  always turning

Elul 25

Shemittah Seder

Nina Beth Cardin

Ever since the first breath of creation, time has unfolded in cycles of seven. Six days reach their crescendo in the seventh day, Shabbat—the Sabbath, the day of rest. Six years reach their crescendo in the seventh year, *Shemittah*—the sabbatical, the year of renewal. Seven cycles of seven years reach their crescendo in the Jubilee year—the ultimate enactment of re-creation.

All three call forth nostalgic images of Eden, when humanity lived in abundance, peace, equity and ease. All offer a way of partial return. But there are differences among them: Jubilee is more fantasy than experience, more vision than practice. And while it remains part of our sacred narrative, it has nonetheless fallen out of our sacred calendar.

Shabbat, on the other hand, is a constant presence. It is celebrated weekly, as time apart, 25-hours of a lived-dream dimension. We enter Shabbat by leaving the work-a-day world and cross into a domain that is edenic, "a taste of the world to come." We are at leisure, eat well, avoid strife and pretend to create one world, diminishing the boundaries that daily divide us.

Shemittah sits between these two. Neither a fantasy nor a constant presence, it is both a vision of a new reality and a practice to be lived in here-and-now. It happens in the same time and space as all other years, only we are to live this year differently, more equitably, more fully, more intentionally than the six years before. It is a year of harmony and celebration with the Earth, when the Land of Israel rests from the agricultural labors imposed upon her yet when she yields sufficient goodness for us all to thrive. It is a year of commonplace *manna*, when food is ours for the taking, but modestly, temperately, with a deep sense of gratitude and awareness; when debts are forgiven and there is equity for all; when property boundaries are suspended and all becomes once again part of the Commons. It is, in short, a year of rebooting, recalibration and realigning our assumptions about property, land use, economic justice and social equity. Not as a dream but as a reality.

This Rosh Hashanah seder is built around the seven-year cycle and incorporates the images of *shemittah*. It is modeled on the Jewish tradition of

New Year's *simanim*, symbolic foods, like the traditional apples dipped in honey, that represent the blessings we hope will be ours.[1]

The seder consists of six small cups or bowls arrayed on a decorative base plate.

This base plate represents the whole, the sweep of time, the sphere that encompasses and defines every seven-year cycle. For *shemittah* is not just one segregated year, as Shabbat is not one segregated day. It is the year that frames and gives shape to all the other years, both those just past, and those yet to come. Upon this foundation plate rest the six cups or bowls. Together they represent the six attributes that define the essence of the *shemittah* year, and a life lived in goodness, sacred striving and delight.

Slices of apples (and other perennial delicacies of your choice) are arrayed in the center of the base plate. These recall the fruits of Eden that sustained us, and the Tree of Knowledge that launched us on the irresistible human enterprise of curiosity, desire, exploration and pursuits. And it represents the perennial foods (fruits, nuts and berries) that grow on their own during the *shemittah* year and that we gratefully eat at a time when we do not plow, sow, reap or commercially harvest the produce of the field.

The cups should be numbered from one to six, with each year's seder starting with the appropriate cup.

Cup One: Honey representing Sova (Enoughness). *Sova* is the feeling of fullness without being stuffed; of contentment through what was given and not wanting anything more; of maximum satisfaction with minimum consumption and disruption. This first cup is filled with honey. Pass around the cup for all to dip the apples in the honey, say:

"This year, may we know no hunger, either spiritual or physical. May we be as readily sated with the delights of life as this cup is filled by these drops of honey."

Cup Two: Wine (consider fruit wine, including Passion Fruit Wine from Israel or homemade date wine)[2] signifying Hodayah (Gratefulness). *Hodayah* is the feeling of gratitude, of deep satisfaction and elusive peace with what we have received. Wine is the age-old symbol of celebration, an expression of shared gratitude. It takes years for the vineyard to grow and produce grapes and time enough

---

[1] The symbolic foods and their accompanying quotes and prayers may be adapted for the needs and urgencies of each year. The Biblical *shemittah* texts are: Exodus 23:10-11, Leviticus 25:1-7, Leviticus 25:20-22, and Deuteronomy 15:1-6. This seder is meant to be a template to be used and adapted as celebrants desire.

[2] Only wine that includes grapes qualifies for the Kiddush blessing: *borei pri hagafen*, who creates the fruit of the vine. The blessing *"Shehakol nihiyah bed'varo"* is said over fruit wines without a grape base. If the blessing over wine (*Kiddush*) and bread (*Hamotzi*) have already been said at the beginning of the meal, no additional blessings need to be recited over the foods of the seder plate.

for the wine to ferment. On the human side, this requires steadfastness, peace, stability, and longevity; on nature's side, cool and heat and sun and rain and rich soil all in the right amounts—surely things to be grateful for. This cup is filled to the rim with the wine. (Wine cups at everyone's place may be filled with this too.) Hold it up and say:

"May we know peace and be strangers to disappointment and disruption. May the Earth find renewal through our practices. And may gratitude fill us all as the wine fills this cup."

Cup Three: Figs representing Revaya (Abundance). *Revaya* is the awareness of the vast resources of a healthy world, the Earth's ancient capacity of growth and self-renewal, and our call to keep it going. Figs are not like most other fruit crops. The fruits on one tree do not ripen all at once but one by one, each in its own time. They offer abundance without surfeit. This cup is filled with figs (either whole or cut, fresh if available, though dried figs are fine too), speckled and spangled with seeds. Pass it around for all to take from and say:

"May we recognize abundance and know no waste. May we celebrate the vast goodness that lies within even the most modest cache of life; may we reverently receive life's abundance and, like the continuous fruiting of the fig tree, give what we can, at the time that is right."

Cup Four: Raisins representing Hesed (Goodness, Kindness, Generosity). *Hesed* is a response to our gratitude for the varieties of gifts we have received in this world. Having received we are moved to give. Such is the nature of the gift. The raisins heaped in this cup signify the sweet, satisfying substance that can be given even after other extractions of goodness have been taken. They recall the leaves, the juices, the wine, the vinegar, the shade, the wood and delight that are all gifts of the grape. In response to all that we have been given, we are moved to give more. Pass around the cup for all to take from and say:

"May we know no greed. May we recognize the gifts we have received and in return realize the manifold ways of giving that lie within each of us."

Cup Five: Pomegranate representing Poriyut (Fertility). *Poriyut* is the creativity, the dynamism, the fecundity that characterizes the majesty of nature. It is what allows us to eat during this year of fallowness and renewal. It is the dormancy that bursts forth, in the right conditions, inspiring the human gifts of imagination, discovery and awe. This cup is filled with pomegranate seeds, symbols of overflowing fertility. Pass the cup around for everyone to taste and say:

"May we know no barrenness, no emptiness. May this year of material enoughness bring forth overflowing acts of discovery, delight and spiritual bounty."

Cup Six: Dates representing Otzar (The Commons). *Otzar* is Earth's shared resources, owned by none and gifted to all. It is the storehouse of the ages, the fundamentals of life that we all depend upon. It is the stuff of Earth and society, natural and cultural, that we share now in our lifetimes and leave behind for others. Our stories, our knowledge, our goods, our homes, our Earth. This cup holds stuffed dates, signifying all that we share in the giving to and taking from the Commons. (Another option: put a few symbolic dates in the center cup, but in addition, array dates—pitted and sliced—on the outer edge of a serving plate, surrounding a center mound of stuffing: chopped almonds, walnuts, pistachios or pine nuts that have been soaked in honey and wine. Let everyone fill a date with the sweet filling and give it to someone else at the table.) Everyone takes a date and says:

"May we know no isolation, no loneliness, no selfishness. May we recognize that we are joined in partnership to the Earth, and to one another through our common heritage, the Torah, our past and our future that bind us to one another forever, throughout the cycles of space and time."

Then wash it all down with a drink of *l'chaim*—to life.

Confession

Judith Felsen

Tonight I craved Your company
time alone with You.
I ran uphill
to see Your face
extended through the Elul sunset
drenched in colors
Bikkurim.[1]
The sky, Your temple,
quenched desires
sacred space
communion
mountains holding
context, presence.
Prayers sans words expressed
silent lips
heart happy
sky's violet feast
dessert slowly dipping
behind hills;
welcome
Elul Day
delighting Elul sunset.

---

[1] First fruits.

Choosing Again What Is Good

Joelle Novey

Pope Francis issued an encyclical teaching on ecology, Laudato Si, in June 2015.[1] In the year and a half that followed, I had the opportunity to sit with good folks of many faiths to study its words.

Over that same time frame, I had many moments of feeling overwhelmed by the bad things in our world that seemed so much bigger than any one of us: the irrevocable and global suffering already being caused by our damaged climate; the harm being done to black bodies and spirits by the pernicious persistence of racism; the unrelenting meanness of the presidential campaign rhetoric.

What gave me hope as we entered that season of reflection?

I turned to Pope Francis, and to the paragraph (#205) of his encyclical which never fails to give me a jolt of hope:

"Yet all is not lost. Human beings, while capable of the worst, are also capable of rising above themselves, choosing again what is good, and making a new start, despite their mental and social conditioning. We are able to take an honest look at ourselves, to acknowledge our deep dissatisfaction, and to embark on new paths to authentic freedom.

"No system can completely suppress our openness to what is good, true and beautiful, or our God-given ability to respond to His grace at work deep in our hearts. I appeal to everyone throughout the world not to forget this dignity which is ours. No one has the right to take it from us."

Where does the Pope find some hope for all of us? He finds it right in the place where the shofar finds us, where that still, small voice is heard—he finds it in our hearts; he finds it in our own individual capacity to know what is good and to make choices.

The world is full of terrible things much larger than any of us, but we do have the freedom to make choices—and in that freedom lies the possibility of our redemption. We can choose, and so, we can change. And so can everyone else. And so can our world.

We can choose how we get our energy, how we invest our money, what we will buy, what food we will eat. We can choose to examine our own role in the fossil-fueled economy, the part we play in the evil of systemic racism, and, come each election day, we can choose what kinds of leadership to exalt with our votes.

---

[1] Pope Francis. *Laudato Si': On Care for Our Common Home* [Encyclical], 2015.

That fall of 2016, I let Pope Francis set my *kavannah* for the season of repentance. He reminded me that all was not lost, because we can still decide to choose what is good.

The dignity to engage in *tikkun* this Elul is ours. It is our liberation and our highest hope in these times. And no one can take it from us.

Elul 26

Celebrating the *Shemittah* Cycle

Nina Beth Cardin

Do you know where this new year falls in the *shemittah* (seven-year) count? Or when the next *shemittah* year will be?

Even those of us who were deeply engaged in celebrating the last *shemittah* year may have difficulty remembering when exactly it was. (It was 5775, 2014-2015.) Yet *shemittah*, like Shabbat, is more than a single day, a single slice of time. It is a presence, an ever-coming moment that is in a way always with us. It is a practice, an attitude, a social, economic and spiritual ethic that guides our lives.

In the biblical era, this was evident, and the *shemittah* ethic was a constant reality. As weekdays counted up to the celebration of Shabbat, so years counted up to the celebration of *shemittah*. Years One and Two (as they were designated), as well as Years Four and Five, were years when the annual tithe (the gifts of the divine partnership between God, Earth and humans—or the monetary equivalent thereof) were taken to the Temple and enjoyed there. Not just consumed or spent but enjoyed: "Use the silver to buy whatever you like: cattle, sheep, wine or other fermented drink, or anything you wish. Then you and your household shall eat there in the presence of the Lord your God and rejoice." (Deuteronomy 14:26)

Years Three and Six were different. In those years, the gift of the tithes was to stay in the farmer's home community—stored in a publicly accessible spot so that those in need could readily take when and what they needed. The handling of one's harvest demanded that the farmer—and the indigent—know what year it was.

Even more, given that any debts owed would have to be reconciled before, or else be forgiven in the seventh year, anyone on either side of an outstanding loan would know what year it was.

These traditions must have conjured up a deep awareness of shared time. Everyone was immersed in the collective passage of years, knowing what is expected of them this year and next, all leading to the grand pause, the reset, the leveling and renewal of the *shemittah* year itself.

I wanted to feel this presence; to be immersed in this cyclical flow of time of my people, to know what time it was, not just in hours or days, but years. So for this past *shemittah* year, I created a *shemittah* cycle Rosh Hashanah seder plate.

It is designed to be placed on the table the first night of the year, every year, as a visual mnemonic of the *shemittah* cycle. It is a circular platter with six small bowls around one central receptacle—representing the fullness of the *shemittah* cycle. Each year, the bowls are to be filled with foods symbolic of that year, with a different bowl (corresponding to the year of the cycle) being the lead each year. Ideally, the food in the bowls—and the telling that accompanies them—will heighten our awareness of the values and the tasks we are called to do, with an emphasis on our part in renewing all of God's creation.

My effort is a pale version of what I hope can eventually be developed by potters, woodworkers, sculptors and other artists, so that they might craft their own versions to inspire us all.

Resilience in the Face of Adversity

Susie Davidson

As a writaholic, I am also a readaholic. As we move toward creating communities that feed, nurture and sustain all the inhabitants of the Earth, and the Earth itself, I believe that it is also incumbent upon us to remain informed about the news of the day and the topics that affect underlying societal infrastructures.

Certainly, some of these infrastructures seem entrenched to the point of impermeability, none more so than the economic systems that govern world relations. For those of us concerned with environmental health and sustainability, there is possibly no greater challenge.

We embrace *t'shuvah* and serve G-d by returning and adhering to our highest visions. And none can be higher than safeguarding the planet that we live on.

In 2015, Pope Francis released his climate-centered encyclical Laudato Si, which translates to "May the Creator Be Praised" and is taken from a prayer of St. Francis of Assisi acknowledging Brother Sun, Sister Moon, and all other elements of Creation. To enthusiastic worldwide reception, the encyclical stated that humans were morally-bound to protect the planet for future generations and especially for the vulnerable among us.

But the next day, by one deciding vote, the U.S. Senate Appropriations Committee effectively gutted the EPA's first-ever plan to implement limits on carbon pollution from existing power plants. And from then on, things only got worse, to the point where it is easy to even lose hope. But we can't. Instead, we can redouble our efforts.

"We are the first generation to feel the impacts of climate change, and the last generation to be able to do something about it," said then-President Barack Obama.[1]

Who is the Jewish counterpart to the Pope and ecologically active leaders such as President Obama? Where is our Moses, our King David, our David ben-Gurion, to lead us to victory against fossil fuel defenders and enablers?

I nominate Rabbi Arthur Waskow, founder and Director of the Shalom Center in Philadelphia.

Waskow, whose numerous books and writings include *Torah of the Earth: Exploring 4,000 Years of Ecology in Jewish Thought, Trees, Earth, & Torah: A Tu B'Shvat Anthology*, and "Jewish Environmental Ethics: Adam and Adamah" in the *Oxford Handbook of Jewish Ethics and Morality*, is consistently in the forefront of Jewish leadership climate actions, such as the Rabbinic Letter on the Climate Crisis initiated in anticipation of Laudato Si.

Hundreds of rabbis signed onto the Shalom Center's statement against fossil fuel-extracting practices such as fracking, offshore Arctic drilling, and oil trains, and their disproportionate impacts on low-income communities and communities of color. "Climate Pharaohs" is the term Waskow applies to foes of the environment. "'Carbon Pharaohs'...endanger human beings and bring plagues upon the Earth," the rabbis wrote in the Letter.[2]

We can all be grateful to Pope Francis, the past efforts of President Obama, Rabbi Arthur Waskow and his staff at the Shalom Center, and all eco-defenders who act individually, as leaders, and/or as members of organizations to protect and safeguard natural life on Earth, now and for years to come. But in order to do our own *t'shuvah*, to return to G-d, we must do more than be grateful—we must all do the work of defending the Earth.

---

[1] "Step-by-Step Guide to Understanding Obama's Clean Power Plan." *Utilities | Energy Digital*, Energy Digital Staff, 5 Aug. 2015, http://www.energydigital.com/renewable-energy/step-step-guide-understanding-obamas-clean-power-plan

[2] Waskow, Arthur. "Rabbinic Letter on Climate—Torah, Pope, & Crisis Inspire 425+ Rabbis to Call for Vigorous Climate Action." 29 Oct. 2015. Retrieved from theshalomcenter.org/RabbinicLetterClimate

Elul 27

Saving the Earth to Save Our Children

Andy Oram

The traditional Torah readings on Rosh HaShanah cover two of Abraham's most difficult trials, calling on him to relinquish his two sons. The troublesome stories can tell us a lot about how to make room in our lives for our children—and also a lot about how to save the Earth from the devastation of climate change and ecological destruction.

In the first story, read traditionally on the first day of the holiday, Sarah abruptly demands that Ishmael be thrown out with his mother into the desert, and God backs her up (Genesis 21:9-14). Abraham reluctantly goes along. God's approval suggests that Sarah had an understanding not visible on the surface. Abraham is prosperous and can easily support two sons. But for some reason, keeping them in the same tent would not be a long-term solution.

The mythic aspect of this apparently cruel abandonment is revealed in details. Hagar, Ishmael's mother, handles him like a small child even though we know from previous passages that he is a teenager. The literal expulsion into the desert probably did not actually happen. What we do know is that Ishmael survived, because Abraham provisioned him for his journey and because he discovers (despite Hagar's despair) that the desert can sustain him. Ishmael becomes a great nation, returns to honor Abraham at his burial, and even furnishes a wife for one of Abraham's grandchildren.

The progression from the banishment of Ishmael to the binding of Isaac in the following chapter of Genesis illustrates some sort of evolution in the world around them. Perhaps overpopulation had complicated the process of going out to make one's fortune. Canny readers have questions about exactly what God told Abraham to do when it was time to send out Isaac into the world, but it seems that Abraham interpreted the mandate in some horribly distorted way.

My own guess is that God asked Abraham to set up Isaac so he could support himself, and that Abraham did so in a way that could destroy the environment. The modern equivalent is to set up belching factory furnaces that darken the sky with carbon emissions or to sweep down whole forests in order to plant commercial crops. These are the activities that destroy the Earth and our children's future with it. Abraham was doing something, even if done out of love and concern, that would rob Isaac of life.

God realizes Abraham's mistake, in which God may also share some of the blame. So, God gives Abraham a lesson, showing him an alternative way to meet the goal. The ram found by Abraham was on its way to death, caught in a thicket it could not escape. Instead of acting destructively, Abraham carried out what we nowadays would call recycling. He turned the doomed ram into a blessing before God, and Isaac was saved.

Thus, our High Holiday readings warn us to think of the consequences of acts we take on behalf of our children. Every decision we make has long-term effects on our environment, which return to affect future generations. We can take heart in knowing that an element of the divine resides in these seemingly unconnected choices.

Today the World Is Pregnant with Possibility, and So Are We

David Seidenberg

On Rosh Hashanah, we hear the shofar and call out in response, "*Hayom Harat Olam!*" "Today is the birthday of the world; today the world was born."

So says the liturgy, according to most prayer books. This birthday is not just one of celebration but of reflection, as the next line of the liturgy says: "*Hayom ya'amid ba-mishpat.* Today the world stands in judgment." These two motifs would give anyone pause to consider what we are doing to the planet.

But let's look more closely at these words, to see what they really mean and what else they can teach us.

'*Harah*' means pregnancy, conception or gestation. Not birth, but the process which leads to birth. If we wanted to say "the birth of the world," we would say "*leidat ha-olam.*" "*Olam*" can mean "a world," but if we wanted to say "the conception of the world," we would similarly say "*harat ha-olam,*" not "*harat olam.*" "*Olam*" without the prefix "*ha-*" really means "eternity," from the root that means "hidden," or more precisely, the infinite that is hidden and lies beyond our limited perception. So *harat olam* literally means "pregnant with eternity," or "eternally pregnant."

The day of Rosh Hashanah is pregnant with eternity.

What deeper evocation of this wondrous moment and this miraculous Creation could one find than the image of it being "eternally pregnant"— always bringing forth new lives, new creatures, even new species? Always dynamic, growing; balanced not like a pillar on its foundation, but like a gyroscope, turning and turning. What higher praise could there be of the Creator? What greater potential could we recognize in this moment, than to say that it is "pregnant" with insights, with hopes, as great as eternity"?

But we can dig deeper still. In Tanakh, the source of the phrase *"harat olam"* is part of an expression of profound personal grief. Jeremiah says, *"Vat'hi li imi kivri v'rachmah harat olam.* Would that my mother become for me my grave, and her womb be pregnant eternally." (Jeremiah 20:17) Jeremiah was wishing he was never born. In Job, however, our planet is imagined as a womb, as it says, *"yam b'gicho meirechem yeitzei* when the sea gushed forth from the womb." (Job 38:8) When Jeremiah's lament is applied to the Earth, it becomes transformed into one of the truest and most loving sentences in the Tanakh: Let this Earth be a mother to us and our grave; let it be eternally pregnant, so that from our deaths will come new life and new lives.

What about the second line we respond with, *"Hayom ya'amid bamishpat"*? The phrase *"ya'amid bamishpat"* comes from Proverbs: *"Melekh b'mishpat ya'amid aretz.* A king through justice makes the Earth stand." (Proverbs 29:4) *Ya'amid* doesn't mean "stand" but rather, to be caused to stand, to be sustained; the prefix *ba-* can mean "in," but here it means "through, because of." So this response can mean: "This day will be sustained through Justice."

When we hear the shofar and call out, *"Hayom harat olam!"* may we find hope, may we find courage, may we find blessing, in this moment on this planet, filled with birth and death, pregnant with eternity. And may we respond with justice.

Hayom harat olam. "This day births new intentions, conceives new possibilities." This day that is our day, the day we are alive on this planet. As we say at the end of *musaf, "Chayim kulkhem hayom.* All of you are alive today."

Today our choices, our vows, will gestate the future, not just for our children, but for the children of every species upon the Earth. *Hayom t'amtzeinu.* "Today may we find courage." *Hayom t'varcheinu.* "Today may we be blessed." *Hayom ticht'veinu l'chayim tovim.* "Today may we be inscribed to live well"—we and all our relations, all the species who are traveling together on this planet.

Hayom im b'kolo tishma'u. "Today, if you will listen to the Voice."

Elul 28

Eco-Kaddish Blessing

Judith Felsen

> Master of the Universe, Lord of All Worlds,
> may I know You
> when I see a tree
> feel the wind
> gaze at stars
> touch the ground
> smile at flowers
>
> May I see You
> in life's cycle
>
> May I instill You
> with every breath
> footstep, touch, thought
>
> This Elul
> may I join You in
> nature's presence
> never to leave.

Repentance, Prayer, and Deeds of Righteous Action Will Stop Climate Change

Mirele B. Goldsmith

This year, as the sun sets on Yom Kippur, our prayers will reach a pinnacle of intensity as we recite the *Unetaneh Tokef* prayer: "On Rosh Hashanah it is written, and on Yom Kippur it is sealed. How many shall leave this world, and how many shall be born; who shall live and who shall die, who in the fullness of years and who before; who shall perish by fire and who by water, who by sword and who by a wild beast; who by famine and who by thirst…But repentance, prayer, and deeds of righteous action, can remove the severity of the decree."

The *Unetaneh Tokef* was written ages ago, perhaps as early as the first century, but it is eerily contemporary in the way in which it describes the life and death consequences of climate change. Although climate change is a new cause of death, the ways in which human beings are vulnerable, suffer, and die, are timeless. Death comes by water when floods result from devastating storms and rising seas. Death comes by wildfire when drought is worsened by climate change. Death comes by famine when rising temperatures turn farmland into desert.

The solution is in our hands. "Repentance, prayer, and deeds of kindness can remove the severity of the decree." The Gates of Mercy are never closed. It is up to us as human beings to exercise our free will to change the course of history. The call to repentance, prayer, and deeds of righteous action, is a personal challenge to every Jew.

I Can Do *Something*

Joan Rachlin

I once heard it said that most working folk are "denatured," so one of my post-retirement goals has been to "renature." With this *kavannah* in heart and mind, I have been trying to more actively appreciate the boundless gifts nature offers us daily.

Most specifically, I've begun to notice, appreciate, and more consistently support those who produce the food that sustains my family and me. Through the physical labor of farmers, we are given the gift of nourishment, which fuels us as we engage in our chosen pursuits and passions. And through the stewardship of farmers, the Earth receives the gift of care, which enables it to remain healthy and fertile. The farmers I've met love their land, respect their plants and animals, and recognize their synergistic relationship with the Sun, rain, winds, and seasons. They rely on that relationship for their livelihood, but it is now threatened by climate change.

As Edward Everett Hale said, "I cannot do everything, but still I can do something."[1] I joined a CSA (community-supported agriculture) and go to farmers' markets each week. I buy my eggs, honey, and beeswax candles from a farm where the chickens rest peacefully under bushes as though posing for a still life. I buy chicken from a farm family that feeds their animals by hand,

---

[1] Grover, Edwin O. *The Book of Good Cheer : A Little Bundle of Cheery Thoughts*. Chicago: P.F. Volland, 1909, p. 28.

fretting over any that are not doing well. They "dress" those chickens by holding them gently and killing them softly and swiftly. This farmer accompanies his cows to the nearby slaughterhouse the night before they die in order to feed them their last supper; he then sleeps with them in the barn to ensure that they are calm despite the new surroundings.

The food I purchase from farmers feels holy, and wasting it would thus be tantamount to disrespecting them. I therefore compost kitchen scraps, eggshells, coffee grinds, and ash from our grill, being careful not to squash the bugs and worms feasting on my garbage for they, too, are part of the food-waste-to-rich-soil continuum. Everything is connected.

I saw my first red leaf last week with its bittersweet message: summer is turning toward fall and it is thus time for me to enter this season of self-reflection. I have missed the mark by taking farmers and our Earth for granted, but *t'shuvah* (repentance) affords me the perennial gift of intentionality and change. I hope to put my time, money, and mouth where my words are when it comes to honoring and supporting local farmers, sustainable food producers, and vendors and nonprofits working for a just food system.

Among other things, I'll be thinking about how I can contribute to inner-city agriculture. The Urban Farming Institute, for example, runs training programs, provides land access, engages in public education, and produces food for those who are "food insecure." A thriving urban food system combines the elements of *tikkun olam* (healing the world) and *tikkun tevel* (healing the Earth), and I hope to find a way in which I might participate in that sacred work. If we're not careful and caring, successive generations might not have access to nourishing produce during Elul, or at all. That prospect is overwhelmingly sad and frightening, and although I cannot do everything to prevent it, I can do something.

Elul 29

Rocks in my Life

Margaret Frisch Klein

They say that Rosh Hashanah is the birthday of the world. It is an opportunity filled with new beginnings. Everything seems fresh and new. So much more so out in G-d's glorious Creation, singing psalms that express that majesty. Many Rosh Hashanah mornings have found me at Plum Island before sunrise or at Walden Pond trying to figure out, in Thoreau's words, "I went to the woods to learn to live deliberately."[1]

They say that G-d is a Rock, capital R, *Adonai Tzuri*, G-d is My Rock. When I was first learning Hebrew this was the only word I knew for rock or stone. The Israelis laughed when I tried to use it to describe the beautiful Jerusalem stone. *Tzur* is only for G-d, they told me. But sometimes people get closer to G-d sitting on rocks. Jacob uses a stone for a pillow, had a dream and woke up saying, "G-d was in this place and I knew it not." (Genesis 28:16)

Once, I was sitting on the rocks on the Marginal Way in Ogunquit, Maine. In Maine, they even have an expression for this. The original tourists, rusticators, those summer people who came to places like Ogunquit and Bar Harbor by steamer, stagecoach or train, would sit on the rocks for hours just looking at the ocean, thinking or painting. They called it rocking. As I sat there, I was thinking of all the times I have sat there. Many major life decisions have been made sitting on those very rocks. My husband and I decided to have a child sitting there on a cold February morning. One April, I rocked to decide whether I could finish rabbinical school, despite some overwhelming obstacles. One July, I rocked and debated whether to accept a position as an educational director after ordination. We have celebrated Thanksgivings and Mother's Days on those rocks. More recently, I returned to Ogunquit for my birthday all by myself to walk the beach and the Marginal Way, to sit on those rocks and to figure out what my vision of the rabbinate is. I completed my application for Congregation Kneseth Israel in my hotel room that night. I was impressed with their vision process. It seemed to mirror mine. I took the job.

But accepting that position meant leaving those beloved rocks. My last trip to Ogunquit before the move was a bright, sunny day. The ocean was a deep blue against the sky. It was breathtaking. When I stepped out of the car,

---

[1] Thoreau, Henry D. *Walden: or, Life in the Woods*, 1854. VT: Tuttle Publishing, 1995.

I said to myself, "How can I leave this place?" I even called my daughter, then in New York and said I couldn't leave. Then I sat there. I realized that those rocks will be there. They are eternal. I can return to them. Again and again. The High Holiday liturgy says that we can return. Sitting on those rocks helps me prepare. Sitting on those rocks is a real concrete (pun intended) form of *t'shuvah*, return. To the rocks. To the Rock. To sit and meditate again.

They say that Rosh Hashanah is the birthday of the world. Nowhere is that more apparent than where the rocks and the water meet. May we all find a way return.

Bringing Truth to Power

Hattie Nestel

To bring truth to power on any day is always rewarding, but to bring truth to power on Yom Kippur means even more.

I met Philip Berrigan in 1982. Along with his wife, he founded the Atlantic Life Community (ALC), a mixture of Jewish and Catholic activists from the East Coast. ALC activists encouraged me to join them in blocking a Trident nuclear submarine on Yom Kippur, 1983, during a non-violent direct action. My two sons, Kenny, 17, and Gad, 9, enthusiastically joined me.

That a Trident launch occurred on Yom Kippur enhanced the meaning for our little Jewish family. Perhaps twenty of us, including Kenny and me, decided to risk arrest by going under police barricades to block the entrance to the launch. Police arrested us. Gad blew the shofar continuously until the police released us.

To have taken the step of being arrested shocked and liberated me. "Beating swords to plowshares and spears into pruning hooks" (Isaiah 2:4) informed many ALC actions I participated in.

I was born in 1939 to Conservative Jewish parents in a mostly gentile suburb of Philadelphia. I lived my first six years in a household dominated with the knowledge and fear of what was happening to Jews in Nazi Germany.

From the earliest age, I remember the confusion and fear of waking up at night to the sound of American air raid sirens during World War II. My parents drew down all the blackout shades until we heard the all-clear sirens. Fear welled in me when we went to the neighborhood movie theater where there were newsreels about the war. We watched clips of Hitler and saluting, flag waving, cheering mobs of thousands. I knew that Hitler and his Nazis hated Jews, and although I do not remember being told I was a Jew, I just knew it. I have never forgotten those images.

I often overheard my parents whispering about buying guns for Jews in Nazi-occupied countries. They worked in their own way to stop Jewish death and destruction.

During the war, my father planted a backyard Victory garden with a small strawberry patch he gave me to water and pick at the end of the yard. Unfortunately, the strawberry patch abutted our backyard neighbor's house where the gentile children knew we were Jewish. While I picked strawberries, they threw stones at me and called me a Jesus killer. Again, being Jewish made me fearful.

As I lived through young adulthood, I began to understand more deeply what happened during the Holocaust. My mother often retold her remembrances of 1939 when Jewish refugees aboard the *St. Louis* ship were not allowed to dock.

My mother gave me the book, *Blessed Is the Match*,[1] about Hannah Senesh, who resisted the Nazis and was eventually executed for attempting to liberate Hungarian Jews. From then on, I avidly read everything about the Holocaust. I read story after story, history after history of Jewish persecution. Those stories are like cells in my body. They are never far from my mind.

Recently, activism led me to work with many others to stop a fracked gas pipeline proposed to traverse Massachusetts.

I see pipelines as a destroyer of life, another instrument of climate destruction. I resolved to tell the stories of those whose lands would be destroyed by the pipeline. I took a course and slowly learned to use a video camera and edit footage. I completed 37 interviews that aired on 30 Massachusetts community cable access stations. I am resolved that families and land will not be destroyed without a fight. I will not be a bystander. This is my current way of beating swords into plowshares and spears into pruning hooks.

I do not know where I will put my body on Yom Kippur this year, but I know it will be somewhere on the pipeline route praying that it can be stopped.

---

[1] Syrkin, Marie. *Blessed is the match: The story of Jewish resistance.* Jewish Publication Society, 1947.

Biographical Notes

Rabbi Adina Allen is Co-Founder & Creative Director of The Jewish Studio Project (JSP). Innovating an entirely new Jewish environment: part *beit midrash* (house of Jewish learning), part urban art studio, and part spiritual community, JSP utilizes the creative arts as a tool for self-discovery, social change and for inspiring a Judaism that is vibrant, connective and hopeful. Rabbi Allen is a contributing writer to the *Huffington Post.* She's been published by the *CCAR Journal, Sh'ma Now, Kveller, Jew School*, and the *Journal of Interreligious Studies.* Ordained by Hebrew College in 2014, Rabbi Allen is an Open Dor Fellow and an alum of the Upstart Fellowship and the Wexner Graduate Fellowship. She serves on the Board of Directors of Urban Adamah and lives in Berkeley, California. www.jewishstudioproject.org

Rabbi Katy Z. Allen is the founder and rabbi of Ma'yan Tikvah—A Wellspring of Hope, which holds services outdoors all year long. She is the co-founder and President pro-tem of the Jewish Climate Action Network. She is a Board-Certified Chaplain through *Neshama: Association of Jewish Chaplains* and is a former hospital and hospice chaplain. She now serves as an Eco-Chaplain and the Facilitator of the One Earth Collaborative, a program of Open Spirit in Framingham, Massachusetts. She received her ordination from the Academy for Jewish Religion in Yonkers, New York in 2005, and lives in Wayland, Massachusetts. www.mayantikvah.org

Rabbi Steve Altarescu currently serves as co-rabbi with his wife, Rabbi Laurie Levy, at the Reform Temple of Putnam Valley. He was ordained in 2014 at the Academy for Jewish Religion. He holds a Bachelor's degree in Religious Studies and Literature and a Master's degree in Counseling. Rabbi Altarescu is a Board-Certified Chaplain through *Neshama: Association of Jewish Chaplains.*

Tali Anisfeld is a student at Princeton University where she studies Anthropology and Environmental Humanities, employing anthropological methods and theory in order to think more deeply about issues of environmental justice. She is a leader of Princeton's student interfaith organization, bringing our relationship to the natural world into their exploration of spirituality. Anisfeld grew up in Newton, Massachusetts.

Maggid **David Arfa** (Mah-geed; Storyteller) has produced two CD's, "The Birth of Love: Tales for the Days of Awe", and the Parents' Choice Award-winning "The Life and Times of Herschel of Ostropol: The Greatest Prankster Ever To Live." His full-length performance, "The Jar of Tears: A Memorial for the Rebbe of the Warsaw Ghetto," won the Charles Hildebrandt Holocaust Studies Award. *Maggid* Arfa is currently the full-time chaplain at Providence Behavioral Health Hospital in Holyoke, Massachusetts. He also leads Shabbat-inspired contemplative services at the local Audubon High Ledges and weekly Mikvah's in the Deerfield River by his home in the Berkshire foothills. www.maggiddavid.net/

Rachel Aronson is an educator, nature lover, and dialogue facilitator living in Brooklyn. She has worked in the intersection of environmentalism and spirituality at Hazon, Resetting the Table, and Kolot Chayeinu.

Molly Bajgot is a Jewish singer-songwriter and community organizer from Sudbury, Massachusetts. She currently lives in Northampton, Massachusetts. She is a lover of music, healing arts, and the outdoors. When she gets the chance to do these things together, she feels at home. She loves to craft ritual, and to be in community both as a member and as an organizer. She looks forward to molding all these passions together in to a career throughout her life. www.soundcloud.com/mollybajgot www.joinforjustice.org (Jewish fellowship for folks in their 20's and 30's)

Hazzan Shoshana Brown serves as cantor and co-spiritual leader (along with her husband, Rabbi Mark Elber) at Temple Beth El, in Fall River, Massachusetts. Hazzan Brown combines her love of singing and spiritual leadership by serving as cantor, and her love of nature and writing by writing monthly hiking articles for the *Fall River Herald News*. She loves that her assignments for the newspaper have made her get out in nature and also led her to learn a great deal about the unique ecosystems of Southeastern Massachusetts and Rhode Island. Recently, Hazzan Brown has added nature photography to her satchel. www.frtemple.org

Rabbi Nina Beth Cardin is the co-founder of the Sova Project and Advocacy Chair of the Baltimore Jewish Sustainability Coalition; founder of the Baltimore Orchard Project, an organization that grows, gleans and gives away urban fruit; and a co-founder of the Interfaith Partners for the Chesapeake, an interfaith organization that works on behalf of the health of the Chesapeake Bay watershed and all its inhabitants.

Sarah Chandler holds an M.A. in Jewish Education and an M.A. in Hebrew Bible from the Jewish Theological Seminary, and a certificate in Non-Profit Management and Jewish Communal Leadership from Columbia University. She teaches, writes and consults on issues related to Judaism, Earth-based spiritual practice, the environment, mindfulness, food values, and farming. She has served as Director of Earth Based Spiritual Practice for Hazon's Adamah Farm. Recently, she was the CCO (Chief Compassion Officer) at JIFA (Jewish Initiative For Animals). Ordained as a Kohenet (Hebrew Priestess) in 2015, she is studying as a shamanic healer apprentice at The Wisdom School of S.O.P.H.I.A. and Kabbalistic imaginal dream work at The School of Images. She lives in Brooklyn, New York.

Rabbi Howard A. Cohen serves Congregation Shirat Hayam on Boston's south shore. In addition, he is the founder, CEO and senior guide of Burning Bush Adventures: Judaism Outdoors, the CEO of Pemsikwa Life Coaching and a Captain in the Bennington Fire Department, Bennington, Vermont. www.rabbihowardacohen.com

Rabbi Michael M. Cohen is a founding faculty member of the Arava Institute for Environmental Studies where for over twenty years it has worked to prepare future leaders from Israel, Palestine, Jordan, and around the world to cooperatively solve the regional and global challenges of our time. He also teaches Conflict Resolution at the

Center for the Advancement of Public Action of Bennington College and is Rabbi Emeritus of the Congregation Manchester Center, Vermont. www.arava.org

Sandra N. Daitch lives and works in Arlington, Massachusetts, where she has a small private practice as a Licensed Massage Therapist and Certified Somatic Experiencing Practitioner. She offers Massage classes and private instruction for individuals and couples and teaches Infant Massage. Additionally, Daitch is a Certified Laughter Yoga Leader and enjoys leading laughter events in the Greater Boston area. Sandra's many interests include organic gardening, walking in nature, doing crafts, singing, and dancing, especially English Country, Sacred Circle and Authentic Movement.

Rabbi Robin Damsky is the founder and executive director of In the Gardens, a nonprofit bringing edible organic garden design as well as mindfulness and meditation practice to people in schools, communal organizations, businesses and homes. She serves as the rabbi of Temple Israel in Miller Beach, Indiana. She is a graduate of the Institute for Jewish Spirituality's Clergy Leadership Program and their Jewish Mindfulness Meditation Teacher Training program. Rabbi Damsky was ordained by the Ziegler School of Rabbinic Studies and earned her Master's degree in Jewish education at the Jewish Theological Seminary. She has a BFA in dance from Ohio University and has been a medical massage therapist since 1977. inthegardens.org

Susie Davidson is a poet, journalist, author, and filmmaker who contributes to the *Huffington Post*, *Jewish Daily Forward*, *Wickedlocal.com*, *Jewish Journal* and other media. She is Coordinator of the Boston chapter of the Coalition on the Environment and Jewish Life and a board member of the Alliance for a Healthy Tomorrow and the Jewish Alliance for Law and Social Action's Climate Justice Team. www.SusieD.com

Janna Diamond is a somatic therapist, healer, and activist weaving the wisdom of the body, earth and spirit to guide personal and collective transformation. To learn more, visit: www.jannadiamond.com

Rabbi Dorit Edut is a Jewish educator who was ordained in 2006 at the Academy for Jewish Religion, a pluralistic Jewish seminary in Yonkers, New York. After serving two congregations, Rabbi Edut brought together a diverse group of clergy and civic leaders to find ways to help revitalize the City of Detroit with a focus on its youth, resulting in the Detroit Interfaith Outreach Network (DION). DION's programs partner with inner-city schools to offer literacy tutoring, career exploration and conflict resolution programs, and sponsors drives for books, clothing and sanitary supplies and tours of various ethnic and faith-based centers in the Detroit area. She continues private teaching, speaking, and leading life-cycle events.

Rabbi Jeff Foust is Jewish Adviser and member of the Spiritual Life Center at Bentley University, Waltham, Massachusetts. He is a student and teacher of Kabbalah (embodied spirituality), emphasizing the integration of the spiritual, intellectual, emotional, and material aspects of our lives. He is also active in building positive interfaith relations, in pastoral care and counseling, leading creative services and lifecycle events, and teaching a wide variety of subjects such as Profits and Prophets, Biblical Pathfinders; Davenology (*How to Make Prayer Real*); Kabbalah, meditation, and creative movement. www.rabbijeffreyfoust.com

Judith Felsen holds a Ph.D. in Clinical Psychology, certificates in hypnotherapy, NLP, Eriksonian Hypnosis, and Sacred Plant Medicine. She is a poet, consultant, creator of collaborative integrative programs involving nature, Judaism, spirituality and the arts, student of Torah, sacred texts and various teachers, sacred circle dancer, avid kitchen worker, student of nutrition and volunteer. She enjoys continuous learning and sharing studies with others, the outdoors, hiking, running, meditating and conversing with the Earth. She serves on the board of the Bethlehem Hebrew Congregation, Neskaya Center for Movement Arts, and the Mount Washington Valley Chavurah. She lives in a cabin in the White Mountains of New Hampshire.

Rabbi Moshe Givental is a lover of G-d, people, animals, and all living things. He is a psychotherapist, rabbi, and activist, dedicated to deep listening, the healing potential of relationships, and the transformative power of difficult conversations.

Rabbi Laurie Gold is the rabbi of Temple Beth Elohim in Brewster, New York. She has extensive experience providing pastoral care and has worked at many institutions, including Amsterdam House, The Jewish Home and Hospital, and Chapin Home for the Aging. Rabbi Gold has served as a guest lecturer and service leader at a variety of congregations, including Temple Beth El of Great Neck and Kolot Chayeinu of Brooklyn. She has been the rabbi on cruises for Holland American, Celebrity and Seabourn. Rabbi Gold has taught at Services and Advocacy for GLBT Elders (SAGE), Bronx House, and The Samuel Field Y.

Dr. Mirele B. Goldsmith is an environmental psychologist, educator, and activist. Mirele created the Tikkun Mayim, a ceremony of repair for our relationship with water. She directed the Jewish Greening Fellowship, a network of 55 organizations committed to sustainability. She was a leader in the Jewish mobilization for the People's Climate March and a founder of Jewish Climate Action Network New York City. Mirele's writings on Judaism and sustainability have been published in the *Jerusalem Report*, *Jewish Week*, *Forward*, *Shma*, and *Huffington Post*.

Rabbi/Cantor Anne Heath, a member of both the Greater Rhode Island and Massachusetts Boards of Rabbis, has served since 2003 as spiritual leader of Congregation Agudath Achim and the Jewish Community House—a 110-year-old progressive, independent congregation in the heart of Taunton, Massachusetts. She received her rabbinic ordination from the Academy for Jewish Religion in New York in 2007. www.jewishtauton.com

Thea Iberall, PhD, is on the leadership team of the Jewish Climate Action Network and the Green Sanctuary Committee of First Parish Unitarian Universalist Church in Medfield, Massachusetts. She is the author of *The Swallow and the Nightingale*, a novel about a 4,000-year-old secret brought through time by the birds. In this fable, she shows us that the visions of Gandhi, religious mysticism, and Native Americans are a more sustainable solution than the patriarchal system under which we live. Dr. Iberall is an inductee in the International Educators Hall of Fame. www.theaiberall.com

Rabbi David Jaffe is the author of Changing the World from the Inside Out: A Jewish Approach to Personal and Social Change, winner of the 2016 National Jewish Book Award for Contemporary Jewish Life. He is the Founder and Principal of Kirva Consulting which helps individuals and organizations access spiritual wisdom for creating healthy, sustainable relationships and communities. He blogs at rabbidavidjaffe.com.

Daniel Kieval is a musician, naturalist, educator, and spiritual explorer, among other things. He currently lives in the Connecticut River Valley in western Massachusetts.

Rabbi Margaret Frisch Klein is the rabbi of Congregation Kneseth Israel in Elgin, Illinois. She blogs as the Energizer Rabbi. Most Shabbat afternoons find her outdoors in nature, hiking, running or walking. She partners with many civic organizations to make the world a better place. She honed her love for the water at Mayyim Hayyim, the Community Mikveh in Newton, Massachusetts, where she served as a mikveh guide and educator. She received her rabbinic ordination from the Academy for Jewish Religion in New York. www.theenergizerrabbi.org

David Krantz is the co-founder and president of Aytzim: Ecological Judaism and serves on the board of directors of Interfaith Moral Action on Climate as well as Arizona Interfaith Power & Light. He is a National Science Foundation IGERT doctoral researcher and Wrigley Fellow at Arizona State University's School of Sustainability.

Rabbi Judith Kummer is the Executive Director of the Jewish Chaplaincy Council of Massachusetts. A Boston native, she earned a BA from Barnard College in Environmental Studies and Urban Planning and was ordained at the Reconstructionist Rabbinical College in 1995. Rabbi Kummer is an avid organic gardener, potter, hiker and social activist.

Dr. Scott Lewis is a science and environmental educator and community instigator living in South Florida. His professional interests include project-based approach to science education, the interaction of culture and cognition, and electronic learning. Dr. Lewis is involved in several community volunteer projects including exploring the impacts of Climate Change and supporting efforts to grow environmentally responsible, fairly produced, delicious food. He's also one of the leaders of the Higher Ground Initiative, an effort to highlight the issue of sea-level rise within the Reform movement.

Maxine Lyons, a retired community educator, is an avid gardener. She is exploring the wonderful resonances between Jewish Mussar and Buddhist mindfulness practices while she enjoys some of her time in spiritual accompaniment with homeless individuals. She has been an active member of CMM (Cooperative Metropolitan Ministries) an interfaith, social justice organization as a board member and co-chair of CMM's Interfaith RUACH Spirituality Programs, a 12-year participant in the ALEPH prison pen pal program ("connecting Jews on the outside with Jews on the inside"), and a member of Temple Beth Zion, Brookline, Massachusetts.

Leora Mallach is the co-founder and director of Beantown Jewish Gardens, building community through experiential food and agriculture education rooted in Jewish text, tradition and culture in the greater Boston area.

Rabbi Natan Margalit received rabbinic ordination at The Jerusalem Seminary in 1990 and earned a Ph.D. in Talmud from U.C. Berkeley in 2001. He has taught at Bard College, the Reconstructionist Rabbinical College, the Rabbinical School of Hebrew College and is currently the Director of the Rabbinic Texts department for the ALEPH Ordination Program of Jewish Renewal. Margalit is Rabbi of The Greater Washington Coalition for Jewish Life. He is Founder and President of Organic Torah Institute, a non-profit organization which fosters holistic thinking about Judaism, environment and society (www.organictorah.org). He lives in Newton, Massachusetts. www.organictorah.org

Hattie Nestel is a longtime civil resister, peace walker, and organic gardener who works to save the world by defying war, weapons, and environmental outrages. She currently focuses on stopping pipelines that carry fracked gas. She lives in Athol, Massachusetts.

Joelle Novey directs Interfaith Power & Light (DC.MD.NoVA), which engages hundreds of congregations of all faiths from across the DC area and Maryland in saving energy, going green, and responding to climate change: www.IPLdmv.org. She lives in Silver Spring, Maryland, and davens at Tikkun Leil Shabbat and Minyan Segulah.

Andy Oram is a writer and editor at O'Reilly Media, a technology publisher and conference provider. A member and past president of Temple Shir Tikvah in Winchester, Massachusetts, he is currently interim secretary of the Jewish Climate Action Network. https://www.oreilly.com/pub/au/36

Joan Rachlin is the executive director emerita of Public Responsibility in Medicine and Research, an international bioethics organization. She has also practiced health, criminal, and civil rights law. Joan has been involved with the Women's Health organization, Our Bodies Ourselves, for over 40 years and chaired its Board from 2016-2017. An active member of Temple Israel, Boston, she serves on the Leadership Council, TI Cares, and chairs the Green Team. She received a Distinguished Service Award from the Association of American Medical Colleges in 2013 and the Harvey M. Meyerhoff Award for Leadership in Bioethics from the Berman Institute for Bioethics at Johns Hopkins University in 2014. She holds a J.D. from the Suffolk Law School, and a M.P.H. from the Harvard School of Public Health.

Carol C. Reiman is interested in most things and she tends toward the solitary but finds nature and people engaging. For many years, she has worked with library materials and pets. In the last decade, she has seen major changes, finding comfort in a variety of spiritual traditions.

Leslie Rosenblatt is a wife, a mother, a grandmother. She is a registered nurse and patient advocate. She is a lover of nature and can be found outdoors most days, observing and enjoying nature.

Lois Rosenthal is a member of Temple Tifereth Israel Winthrop. In addition to participating in lay-led Shabbat services, she teaches Hebrew School, prepares students for Bat/Bar Mitzvah, and regularly gives Divrei Torah. She is a member of the local CREW poetry group. She is retired from an academic career in the field of chemistry.

Richard H. Schwartz, PhD, is the author of Judaism and Vegetarianism, Judaism and Global Survival, Who Stole My Religion? Revitalizing Judaism and Applying Jewish Values to Help Heal our Imperiled Planet, and Mathematics and Global Survival, and over 250 articles and 25 podcasts. He is President Emeritus of Jewish Veg, formerly known as Jewish Vegetarians of North America (JVNA), and president of the Society of Ethical and Religious Vegetarians. In 1987, he was selected as Jewish Vegetarian of the Year by JVNA. In 2005, he was inaugurated into the North American Vegetarian Society's Hall of Fame. www.JewishVeg.org/schwartz

Rabbi David Seidenberg is the creator of neohasid.org and the author of Kabbalah and Ecology: God's Image in the More-Than-Human World (Cambridge, 2015), now in paperback. He lives in Northampton, Massachusetts.

Rabbi Jacob Siegel serves as the Director of Engagement for JLens Investor Network, a network for values-based impact investing. Rabbi Siegel received his ordination from Yeshivat Chovevei Torah, a modern Orthodox rabbinical school in New York, and spent a year at Israel's Pardes Institute. He has worked as an educator for Hazon, Eden Village Camp, The 92nd Street Y, the Jewish Farm School. Rabbi Siegel is certified as a shochet (kosher butcher) and has served as scholar-in-residence at colleges and synagogues across the country on Judaism, kosher meat, and ethical food systems, as well as consulting on Jewish sustainable agriculture projects. He is also a certified mohel and serves the community of Southwestern Oregon and Northern California. He lives in Eugene, Oregon.

Howard Smith is a Senior Astrophysicist at the Harvard-Smithsonian Center and the author of *Let There Be Light: Modern Cosmology and Kabbalah, a New Conversation between Science and Religion.* He lectures and writes about science and religion. He lives in Newton, Massachusetts and davens at the Newton Centre Minyan and Shaarei Tefilla. www.LetThereBeLightBook.com

Rabba Kaya Stern-Kaufman, MSW, is the spiritual leader of the Rutland Jewish Center, in Rutland, Vermont. She is a former psychotherapist and feng shui practitioner with a life-long interest in comparative religion and the creation of sacred space and sacred time.

Nyanna Susan Tobin is an Organic Storyteller, Member of Wayland Transition, loves real food, and can often be found walking with her dog and Wellness Partner, Ziggy.

Rabbi Lawrence Troster is the rabbi of Kesher Israel Congregation in West Chester, PA, and he is the Rabbi-in-Residence at the Thomas Berry Forum for Ecological Dialogue at Iona College. Rabbi Troster was the Rabbinic Scholar-in-Residence of GreenFaith, the interfaith environmental coalition in New Jersey and the former creator and director of the GreenFaith's Fellowship program. He is the author of Mekor Hayyim: A Source Book on Water and Judaism.

Rabbi Judy Weiss lives in Brookline, Massachusetts, and is a volunteer climate-change advocate with Citizens' Climate Lobby, Boston Jewish Climate Action Network, and Elders Climate Action's chapter.

Steph Zabel, MSc, is an herbalist and educator who helps urban dwellers connect with the plant world. Through her work, she offers seasonal herbal classes and plant walks. As the founder of Herbstalk—a community-based herbal event in Somerville, Massachusetts—she helps create accessible educational opportunities for all plant enthusiasts. www.flowerfolkherbs.com

Rabbi Ziona Zelazo was ordained at The Academy for Jewish Religion (AJR) in New York. She is the daughter of Holocaust survivors and was born and raised in Haifa, Israel. She served in The Israeli Defense Force and studied Biblical archaeology and ancient languages in Tel-Aviv, Israel. Rabbi Zelazo completed her academic education in Cultural Anthropology at the University of Wisconsin-Milwaukee and taught as an adjunct professor at Montclair State University, NJ. A hospital chaplain, she has served on the bio-ethic committees of Valley Hospital and St. Joseph's-Wayne Hospital. She is a strong advocate for the Women of the Wall in Israel. www.rabbi-ziona.com

Theme Index

Rituals

Turning and returning

Author Index

CPSIA information can be obtained
at www.ICGtesting.com
Printed in the USA
BVHW072054301018
531690BV00001B/22/P